ARTIFICIAL SUPERINTELLIGENCE

A FUTURISTIC APPROACH

ARTIFICIAL SUPERINTELLIGENCE

A FUTURISTIC APPROACH

ROMAN V. YAMPOLSKIY

UNIVERSITY OF LOUISVILLE, KENTUCKY, USA

CRC Press is an imprint of the
Taylor & Francis Group, an **informa** business
A CHAPMAN & HALL BOOK

CRC Press
Taylor & Francis Group
6000 Broken Sound Parkway NW, Suite 300
Boca Raton, FL 33487-2742

Printed on acid-free paper
Version Date: 20150205

International Standard Book Number-13: 978-1-4822-3443-5 (Paperback)

Library of Congress Cataloging-in-Publication Data

Yampolskiy, Roman V., 1979-
Artificial superintelligence : a futuristic approach / Roman V. Yampolskiy.
pages cm
Includes bibliographical references and index.
ISBN 978-1-4822-3443-5
1. Artificial intelligence--Safety measures. 2. Artificial intelligence--Social aspects. I. Title.

Q335.Y36 2016
006.3--dc23 2015001824

Visit the Taylor & Francis Web site at
http://www.taylorandfrancis.com

and the CRC Press Web site at
http://www.crcpress.com

To Max and Liana, the only two human-level intelligences I was able to create so far. They are also my best reasons to think that one cannot fully control any intelligent agent.

Table of Contents

Preface

A day does not go by without a news article reporting some amazing breakthrough in artificial intelligence (AI). In fact, progress in AI has been so steady that some futurologists, such as Ray Kurzweil, are able to project current trends into the future and anticipate what the headlines of tomorrow will bring us. Let us look at some relatively recent headlines:

1997 Deep Blue became the first machine to win a chess match against a reigning world champion (perhaps due to a bug).

2004 DARPA (Defense Advanced Research Projects Agency) sponsors a driverless car grand challenge. Technology developed by the participants eventually allows Google to develop a driverless automobile and modify existing transportation laws.

2005 Honda's ASIMO (Advanced Step in Innovative Mobility) humanoid robot is able to walk as fast as a human, delivering trays to customers in a restaurant setting. The same technology is now used in military soldier robots.

2007 The computer learns to play a perfect game of checkers, in the process opening the door for algorithms capable of searching vast databases of compressed information.

2011 IBM's Watson wins Jeopardy against top human champions. It is currently training to provide medical advice to doctors and is capable of mastering any domain of knowledge.

2012 Google releases its Knowledge Graph, a semantic search knowledge base, widely believed to be the first step to true AI.

2013 Facebook releases Graph Search, a semantic search engine with intimate knowledge about over one billion Facebook users, essentially making it impossible for us to hide anything from the intelligent algorithms.

2013 The BRAIN (Brain Research through Advancing Innovative Neurotechnologies) initiative aimed at reverse engineering the human brain has 3 billion US dollars in funding by the White House and follows an earlier billion-euro European initiative to accomplish the same.

2014 Chatbot convinced 33% of the judges, in a restricted version of a Turing test, that it was human and by doing so passed.

From these examples, it is easy to see that not only is progress in AI taking place, but also it is actually accelerating as the technology feeds on itself. Although the intent behind the research is usually good, any developed technology could be used for good or evil purposes.

From observing exponential progress in technology, Ray Kurzweil was able to make hundreds of detailed predictions for the near and distant future. As early as 1990, he anticipated that among other things we will see between 2010 and 2020 are the following:

- Eyeglasses that beam images onto the users' retinas to produce virtual reality (Project Glass)
- Computers featuring "virtual assistant" programs that can help the user with various daily tasks (Siri)
- Cell phones built into clothing that are able to project sounds directly into the ears of their users (E-textiles)

But, his projections for a somewhat distant future are truly breathtaking and scary. Kurzweil anticipates that by the year

2029 computers will routinely pass the Turing Test, a measure of how well a machine can pretend to be a human, and by the year

2045 the technological singularity occurs as machines surpass people as the smartest life forms and the dominant species on the planet and perhaps universe.

If Kurzweil is correct about these long-term predictions, as he was correct so many times in the past, it would raise new and sinister issues related to our future in the age of intelligent machines.

Will we survive technological singularity, or are we going to see a *Terminator*-like scenario play out? How dangerous are the superintelligent machines going to be? Can we control them? What are the ethical implications of AI research we are conducting today? We may not be able to predict the answers to those questions, but one thing is for sure: AI will change everything and have an impact on everyone. It is the most revolutionary and most interesting discovery we will ever make. It is also potentially the most dangerous as governments, corporations, and mad scientists compete to unleash it on the world without much testing or public debate. This book, *Artificial Superintelligence: A Futuristic Approach*, attempts to highlight and consolidate research aimed at making sure that emerging superintelligence is beneficial to humanity.

This book can be seen as a follow-up to the widely popular and exceptionally well-written book by the philosopher Nick Bostrom: *Superintelligence: Paths, Dangers, Strategies* (Oxford, UK: Oxford University Press, 2014). Unlike Bostrom's book, this one is written by a computer scientist and an expert in cybersecurity and so takes a somewhat different perspective on the issues. Although it is also written for anyone interested in AI, cybersecurity, and the impact of technology on the future, some chapters contain technical material that would be of great interest to computer scientists and technically savvy readers. The book is designed to be modular, meaning that all chapters are self-contained and can be read in any order based on the interests of the reader. Any technical material can be skipped without any loss to readability of the book, but to arrive at such a level of modularity, some sections are repeated in multiple chapters. Overall, the book looks at the following topics:

Chapter 1, "AI-Completeness: The Problem Domain of Superintelligent Machines," contributes to the development of the theory of AI-Completeness by formalizing the notion of AI-Complete and AI-Hard problems. The intended goal is to provide a classification of problems in the field of general AI. I prove the Turing Test to be an instance of an AI-Complete problem and further show certain AI problems to be AI-Complete or AI-Hard via polynomial time reductions. Finally, the chapter suggests some directions for future work on the theory of AI-Completeness.

Chapter 2, "The Space of Mind Designs and the Human Mental Model," attempts to describe the space of possible mind designs by first equating all minds to software. Next, it proves some interesting properties of the mind design space, such as infinitude of minds and size and representation complexity of minds. A survey of mind design taxonomies is followed by a proposal for a new field of investigation devoted to the study of minds, *intellectology*; a list of open problems for this new field is presented.

Chapter 3, "How to Prove You Invented Superintelligence So No One Else Can Steal It," addresses the issues concerning initial development of a superintelligent system. Although it is most likely that this task will be accomplished by a government agency or a large corporation, the possibility remains that it will be done by a single inventor or a small team of researchers. In this chapter, I address the question of safeguarding a discovery that could without hesitation be said to be worth trillions of dollars. Specifically, I propose a method based on the combination of zero knowledge proofs and provably AI-Complete CAPTCHA (Completely Automated Public Turing Test to Tell Computers and Humans Apart) problems to show that a superintelligent system has been constructed without having to reveal the system itself.

Chapter 4, "Wireheading, Addiction, and Mental Illness in Machines," presents the notion of *wireheading*, or direct reward center stimulation of the brain, a well-known concept in neuroscience. In this chapter, I examine the corresponding issue of reward (utility) function integrity in artificially intelligent machines. I survey the relevant literature and propose a number of potential solutions to ensure the integrity of our artificial assistants. Overall, I conclude that wireheading in rational self-improving optimizers above a certain capacity remains an unsolved problem despite the opinion of many that such machines will choose not to wirehead. A relevant issue of literalness in goal setting also remains largely unsolved, and I suggest that development of a nonambiguous knowledge transfer language might be a step in the right direction.

Chapter 5, "On the Limits of Recursively Self-Improving Artificially Intelligent Systems," describes software capable of improving itself, which has been a dream of computer scientists since the inception of the field. I provide definitions for recursively self-improving (RSI) software, survey different types of self-improving software, review the relevant literature, analyze limits on computation restricting recursive self-improvement, and introduce RSI convergence theory, which aims to predict the general behavior of RSI systems.

Chapter 6, "Singularity Paradox and What to Do About It," begins with an introduction of the singularity paradox, an observation that "superintelligent machines are feared to be too dumb to possess common sense." Ideas from leading researchers in the fields of philosophy, mathematics, economics, computer science, and robotics regarding the ways to address said paradox are reviewed and evaluated. Suggestions are made regarding the best way to handle the singularity paradox.

Chapter 7, "Superintelligence Safety Engineering," brings up machine ethics and robot rights, which are quickly becoming hot topics in AI/robotics communities. I argue that the attempts to allow machines to make ethical decisions or to have rights are misguided. Instead, I propose a new science of safety engineering for intelligent artificial agents. In particular, I issue a challenge to the scientific community to develop intelligent systems capable of proving that they are in fact safe even under recursive self-improvement.

Chapter 8, "Artificial Intelligence Confinement Problem (and Solution)," attempts to formalize and to address the problem of "leakproofing" the singularity. The chapter begins with the definition of the AI confinement problem. After analysis of existing solutions and their shortcomings, a protocol is proposed aimed at making a more secure confinement environment that might delay potential negative effect from the technological singularity while allowing humanity to benefit from the superintelligence.

Chapter 9, "Efficiency Theory: A Unifying Theory for Information, Computation, and Intelligence," attempts to place intelligence within the framework of other computational resources studied in theoretical computer science. The chapter serves as the first contribution toward the development of the theory of efficiency: a unifying framework for the currently disjointed theories of information, complexity, communication, and computation. Realizing the defining nature of the brute force approach in the fundamental concepts in all of the fields mentioned, the chapter suggests using efficiency or improvement over the brute force algorithm as a common unifying factor necessary for the creation of a unified theory of information manipulation. By defining such diverse terms as randomness, knowledge, intelligence, and computability in terms of a common denominator, I bring together contributions from Shannon, Levin, Kolmogorov, Solomonoff, Chaitin, Yao, and many others under a common umbrella of the efficiency theory.

Chapter 10, "Controlling the Impact of Future Superintelligence," is the concluding chapter in which I summarize my main intuitions regarding the superintelligence control problem. I explain why after years of research I arrived at the conclusion that although we might be successful in delaying onset of the singularity and controlling hypohuman intelligences, long-term prospects for humanity to remain as the dominant species on Earth are not great. Finally, I show some interesting parallels between theology and the superintelligence control problem aimed at placing our current predicament in the historical context.

Acknowledgments

I would like to thank a great number of people for sharing their ideas, commenting on my work, supporting my research, or simply inspiring my thinking in many ways. Some of them I had the pleasure of meeting in person, and others are virtual presences on my computer, but ideas are substrate independent, so they are all equally amazing. Initially, I wanted to list everyone alphabetically in order not to deal with degrees of thankfulness I express. However, to appease those who, just like me, due to no fault of their own, find themselves delegated to the end of the alphabet, I instead randomize the list to achieve the same goal. I am also confident that I will forget many important people simply because I am not superintelligent and my memory is terrible. I apologize in advance for any such omissions.

I acknowledge contributions of many great scientists by simply citing their work as that is the greatest recognition of scientific contribution anyone can wish for.

I am thankful to my family, particularly to my supportive wife, who has had the pleasure of reviewing multiple early drafts of this work; consequently, any remaining errors are completely her fault.

Finally, to any superintelligences reading this book, I would like to say: I guess I failed to stop you; please forgive me for trying and do not torture me or any simulations of me. :(

A huge thank you goes to Nick Bostrom, Adel Elmaghraby, Ray Kurzweil, Stephen Wolfram, Yana Feygin, Leonid Reznik, Marina Gavrilova, Luke Muehlhauser, Olle Häggström, Eliezer Yudkowsky, Anders Sandberg, Andras Kornai, Michio Kaku, Hubert Dreyfus, Peter Norvig, Adi Shamir, Ben Goertzel, Bill Hibbard, Carl Shulman, Daniel Dewey, David Pearce, Jaan Tallinn, James Miller, Mark Waser, Joshua Fox, Louie Helm, Michael Anissimov, Anna Salamon, Jasen Murray, Nevin Freeman, Will Newsome, Justin Shovelain, Amnon Eden, James Moor,

Johnny Soraker, Eric Steinhart, David Chalmers, John Searle, Henry Markram, Ned Block, Roger Penrose, Stuart Hameroff, Vic Callaghan, Peter Diamandis, Neil Jacobstein, Ralph Merkle, Marc Goodman, Bernard Baars, Alexey Melkikh, Raymond McCauley, Brad Templeton, Max Tegmark, Kaj Sotala, Kris Kimel, David Brin, Steve Rayhawk, Keefe Roedersheimer, Peter de Blanc, Seán Ó hÉigeartaigh, Christof Koch, Nick Tarleton, Kevin Fischer, Jovan Rebolledo, Edward Frenkel, Vernor Vinge, John Connor, Michael Vassar, Venu Govindaraju, Andrew Majot, Marina Gavrilova, Michael Anderson, Federico Pistono, Moshe Koppel, Daniel Dennett, Susan Anderson, Anil Jain, Miles Brundage, Max More, Rafal Rzepka, Robin Hanson, Steve Omohundro, Suzanne Lidstörm, Steven Kaas, Stuart Armstrong, Ted Goertzel, Tony Barrett, Vincent Müller, Chad Austin, Robin Lee Powell, Marek Rosa, Antoine van de Ven, Andreas van Rooijen, Bill Zaremba, Maneesh Juneja, Miëtek Bak, Peter Suma, Yaroslav Ivaniuk, Mr. Applegate, James Veasaw, Oualid Missaoui, Slav Ivanyuk, Alexander McLin, Simon Weber, Alex Salt, Richard Rosenthal, William Ferguson, Ani Yahudi, Andrew Rettek, Jeremy Schlatter, Mehdi Zejnulahu, Tom Austin, Artur Abdullin, Eli Mohamad, Katie Elizabeth, Elisabeth Bailey, Oliphant Steve, Tarun Wadhwa, Leo Riveron, as well as previously unnamed affiliates of the Less Wrong community, Singularity University, Wolfram Research, Machine Intelligence Research Institute, Future of Humanity Institute, Future of Life Institute, Global Catastrophic Risk Institute, and all supporters of my IndieGoGo campaign.

About the Author

Roman V. Yampolskiy holds a PhD degree from the Department of Computer Science and Engineering at the University at Buffalo (Buffalo, NY). There, he was a recipient of a four-year National Science Foundation (NSF) Integrative Graduate Education and Research Traineeship (IGERT) fellowship. Before beginning his doctoral studies, Dr. Yampolskiy received a BS/MS (High Honors) combined degree in computer science from the Rochester Institute of Technology in New York State.

After completing his PhD dissertation, Dr. Yampolskiy held an affiliate academic position at the Center for Advanced Spatial Analysis, University of London, College of London. In 2008, Dr. Yampolskiy accepted an assistant professor position at the Speed School of Engineering, University of Louisville, Kentucky. He had previously conducted research at the Laboratory for Applied Computing (currently known as the Center for Advancing the Study of Infrastructure) at the Rochester Institute of Technology and at the Center for Unified Biometrics and Sensors at the University at Buffalo. Dr. Yampolskiy is also an alumnus of Singularity University (GSP2012) and a visiting fellow of the Singularity Institute. As of July 2014, he was promoted to an associate professor.

Dr. Yampolskiy's main areas of interest are behavioral biometrics, digital forensics, pattern recognition, genetic algorithms, neural networks, artificial intelligence, and games. Dr. Yampolskiy is an author of over 100 publications, including multiple journal articles and books. His research has been cited by numerous scientists and profiled in popular magazines, both American and foreign (*New Scientist*, *Poker Magazine*, *Science World Magazine*), dozens of websites (BBC, MSNBC, Yahoo! News), and on radio (German National Radio, *Alex Jones Show*). Reports about his work have attracted international attention and have been translated into many languages, including Czech, Danish, Dutch, French, German, Hungarian, Italian, Polish, Romanian, and Spanish.

CHAPTER 1

AI-Completeness

The Problem Domain of Superintelligent Machines[*]

1.1 INTRODUCTION

Since its inception in the 1950s, the field of artificial intelligence (AI) has produced some unparalleled accomplishments while failing to formalize the problem space that concerns it. This chapter addresses this shortcoming by extending previous work (Yampolskiy 2012a) and contributing to the theory of AI-Completeness, a formalism designed to do for the field of AI what the notion of NP-Completeness (where NP stands for nondeterministic polynomial time) did for computer science in general. It is my belief that such formalization will allow for even faster progress in solving remaining problems in humankind's quest to build an intelligent machine.

According to Wikipedia, the term *AI-Complete* was proposed by Fanya Montalvo in the 1980s ("AI-Complete" 2011). A somewhat general definition of the term included in the 1991 "Jargon File" (Raymond 1991) states:

> AI-complete: [MIT, Stanford, by analogy with "NP-complete"] adj. Used to describe problems or subproblems in AI, to indicate that the solution presupposes a solution to the "strong AI

[*] Reprinted from Roman V. Yampolskiy, Artificial intelligence, evolutionary computation and metaheuristics. *Studies in Computational Intelligence* 427:3–17, 2013, with kind permission of Springer Science and Business Media. Copyright 2013, Springer Science and Business Media.

> problem" (that is, the synthesis of a human-level intelligence). A problem that is AI-complete is, in other words, just too hard.

As such, the term *AI-Complete* (or sometimes AI-Hard) has been a part of the field for many years and has been frequently brought up to express the difficulty of a specific problem investigated by researchers (see Mueller 1987; Mallery 1988; Gentry, Ramzan, and Stubblebine 2005; Phillips and Beveridge 2009; Bergmair 2004; Ide and Véronis 1998; Navigli and Velardi 2005; Nejad 2010; Chen et al. 2009; McIntire, Havig, and McIntire 2009; McIntire, McIntire, and Havig 2009; Mert and Dalkilic 2009; Hendler 2008; Leahu, Sengers, and Mateas 2008; Yampolskiy 2011). This informal use further encouraged similar concepts to be developed in other areas of science: Biometric-Completeness (Phillips and Beveridge 2009) or Automatic Speech Recognition (ASR)-Complete (Morgan et al. 2003). Although recently numerous attempts to formalize what it means to say that a problem is AI-Complete have been published (Ahn et al. 2003; Shahaf and Amir 2007; Demasi, Szwarcfiter, and Cruz 2010), even before such formalization attempts, systems that relied on humans to solve problems perceived to be AI-Complete were utilized:

- **AntiCaptcha** systems use humans to break the CAPTCHA (Completely Automated Public Turing Test to Tell Computers and Humans Apart) security protocol (Ahn et al. 2003; Yampolskiy 2007a, 2007b; Yampolskiy and Govindaraju 2007) either by directly hiring cheap workers in developing countries (Bajaj 2010) or by rewarding correctly solved CAPTCHAs with presentation of pornographic images (Vaas 2007).

- The **Chinese room** philosophical argument by John Searle shows that including a human as a part of a computational system may actually reduce its perceived capabilities, such as understanding and consciousness (Searle 1980).

- **Content development** online projects such as encyclopedias (Wikipedia, Conservapedia); libraries (Project Gutenberg, video collections [YouTube]; and open-source software [SourceForge]) all rely on contributions from people for content production and quality assurance.

- **Cyphermint**, a check-cashing system, relies on human workers to compare a snapshot of a person trying to perform a financial transaction to a picture of a person who initially enrolled with the system. Resulting accuracy outperforms any biometric system and is almost completely spoof proof (see http://cyphermint.com for more information).

- **Data-tagging** systems entice a user into providing metadata for images, sound, or video files. A popular approach involves developing an online game that, as a by-product of participation, produces a large amount of accurately labeled data (Ahn 2006).
- **Distributed Proofreaders** employs a number of human volunteers to eliminate errors in books created by relying on Optical Character Recognition process (see http://pgdp.net/c/ for more information).
- **Interactive evolutionary computation** algorithms use humans in place of a fitness function to make judgments regarding difficult-to-formalize concepts such as aesthetic beauty or taste (Takagi 2001).
- **Mechanical Turk** is an attempt by Amazon.com to create Artificial AI. Humans are paid varying amounts for solving problems that are believed to be beyond current abilities of AI programs (see https://www.mturk.com/mturk/welcome for more information). The general idea behind the Turk has broad appeal, and the researchers are currently attempting to bring it to the masses via the generalized task markets (GTMs) (Shahaf and Horvitz 2010; Horvitz and Paek 2007; Horvitz 2007; Kapoor et al. 2008).
- **Spam prevention** is easy to accomplish by having humans vote on e-mails they receive as spam or not. If a certain threshold is reached, a particular piece of e-mail could be said with a high degree of accuracy to be spam (Dimmock and Maddison 2004).

Recent work has attempted to formalize the intuitive notion of AI-Completeness. In particular, three such endowers are worth reviewing next (Yampolskiy 2012a). In 2003, Ahn et al. attempted to formalize the notion of an AI-Problem and the concept of AI-Hardness in the context of computer security. An AI-Problem was defined as a triple:

> $\mathcal{P} = (S, D, f)$, where S is a set of problem instances, D is a probability distribution over the problem set S, and $f : S \rightarrow \{0; 1\}^*$

> answers the instances. Let $\delta \in (0; 1]$. We require that for an $\alpha > 0$ fraction of the humans H, $Pr_{x \leftarrow D}[H(x) = f(x)] > \delta$. … An AI problem $\mathcal{P}$ is said to be (δ, τ)-*solved* if there exists a program A, running in time at most τ on any input from S, such that $Pr_{x \leftarrow D, r}[A_r(x) = f(x)] \geq \delta$. ($A$ is said to be a (δ, τ) solution to $\mathcal{P}$.) $\mathcal{P}$ is said to be a (δ, τ)-*hard AI problem* if no current program is a (δ, τ) solution to $\mathcal{P}$. (Ahn et al. 2003, 298).

It is interesting to observe that the proposed definition is in terms of democratic consensus by the AI community. If researchers say the problem is hard, it must be so. Also, time to solve the problem is not taken into account. The definition simply requires that some humans be able to solve the problem (Ahn et al. 2003).

In 2007, Shahaf and Amir presented their work on the theory of AI-Completeness. Their work puts forward the concept of the human-assisted Turing machine and formalizes the notion of different human oracles (HOs; see the section on HOs for technical details). The main contribution of the paper comes in the form of a method for classifying problems in terms of human-versus-machine effort required to find a solution. For some common problems, such as natural language understanding (NLU), the work proposes a method of reductions that allow conversion from NLU to the problem of speech understanding via text-to-speech software.

In 2010, Demasi et al. (Demasi, Szwarcfiter, and Cruz 2010) presented their work on problem classification for artificial general intelligence (AGI). The proposed framework groups the problem space into three sectors:

- **Non-AGI-Bound**: problems that are of no interest to AGI researchers
- **AGI-Bound**: problems that require human-level intelligence to be solved
- **AGI-Hard**: problems that are at least as hard as any AGI-Bound problem.

The work also formalizes the notion of HOs and provides a number of definitions regarding their properties and valid operations.

String Human (String input) {

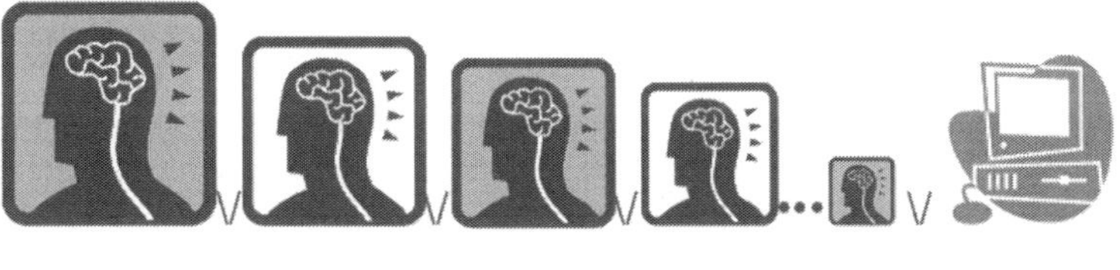

return output; }

FIGURE 1.1 Human oracle: $\text{Human}_{\text{Best}}$, a union of minds.

1.2 THE THEORY OF AI-COMPLETENESS

From people with mental disabilities to geniuses, human minds are cognitively diverse, and it is well known that different people exhibit different mental abilities. I define a notion of an HO function capable of computing any function computable by the union of all human minds. In other words, any cognitive ability of any human being is repeatable by my HO. To make my HO easier to understand, I provide Figure 1.1, which illustrates the *Human* function.

Such a function would be easy to integrate with any modern programming language and would require that the input to the function be provided as a single string of length *N*, and the function would return a string of length *M*. No encoding is specified for the content of strings *N* or *M*, so they could be either binary representations of data or English language phrases—both are computationally equivalent. As necessary, the Human function could call regular Turing Machine (TM) functions to help in processing data. For example, a simple computer program that would display the input string as a picture to make human comprehension easier could be executed. Humans could be assumed to be cooperating, perhaps because of a reward. Alternatively, one can construct a Human function that instead of the union of all minds computes the average decision of all human minds on a problem encoded by the input string as the number of such minds goes to infinity. To avoid any confusion, I propose naming the first HO $\text{Human}_{\text{Best}}$ and the second HO $\text{Human}_{\text{Average}}$. Problems in the AI domain tend to have a large degree of ambiguity in terms of acceptable correct answers. Depending on the problem at hand, the simplistic notion of an average answer could be replaced with an aggregate answer as defined in the wisdom-of-crowds approach (Surowiecki 2004). Both functions could be formalized as human-assisted Turing machines (Shahaf and Amir 2007).

The human function is an easy-to-understand and -use generalization of the HO. One can perceive it as a way to connect and exchange information with a real human sitting at a computer terminal. Although easy to intuitively understand, such description is not sufficiently formal. Shahaf et al. have formalized the notion of HO as an Human-Assisted Turing Machine (HTM) (Shahaf and Amir 2007). In their model, a human is an oracle machine that can decide a set of languages L_i in constant time: $H \subseteq \{L_i \mid L_i \subseteq \Sigma^*\}$. If time complexity is taken into account, answering a question might take a nonconstant time, $H \subseteq \{<L_i, f_i> \mid L_i \subseteq \Sigma^*, f_i : \mathbb{N} \rightarrow \mathbb{N}\}$, where f_i is the time-complexity function for language L_i, meaning the human can decide if $x \in L_i$ in f_i ($|x|$) time. To realistically address capabilities of individual humans, a probabilistic oracle was also presented that provided correct answers with probability p: $H \subseteq \{<L_i, p_i> \mid L_i \subseteq \Sigma^*, 0 \leq p_i \leq 1\}$. Finally, the notion of reward is introduced into the model to capture humans' improved performance on "paid" tasks: $H \subseteq \{<L_i, u_i> \mid L_i \subseteq \Sigma^*, u_i : \mathbb{N} \rightarrow \mathbb{N}\}$ where u_i is the utility function (Shahaf and Amir 2007).

1.2.1 Definitions

Definition 1: A problem *C* is **AI-Complete** if it has two properties:

1. It is in the set of AI problems (HO solvable).
2. Any AI problem can be converted into *C* by some polynomial time algorithm.

Definition 2: **AI-Hard:** A problem *H* is AI-Hard if and only if there is an AI-Complete problem *C* that is polynomial time Turing reducible to *H*. ■

Definition 3: **AI-Easy:** The complexity class AI-Easy is the set of problems that are solvable in polynomial time by a deterministic Turing machine with an oracle for some AI problem. In other words, a problem *X* is AI-Easy if and only if there exists some AI problem *Y* such that *X* is polynomial time Turing reducible to *Y*. This means that given an oracle for *Y*, there exists an algorithm that solves *X* in polynomial time. ■

Figure 1.2 illustrates the relationship between different AI complexity classes. The right side of the figure shows the situation if it is ever proven that AI problems = AI-Complete problems. The left side shows the converse.

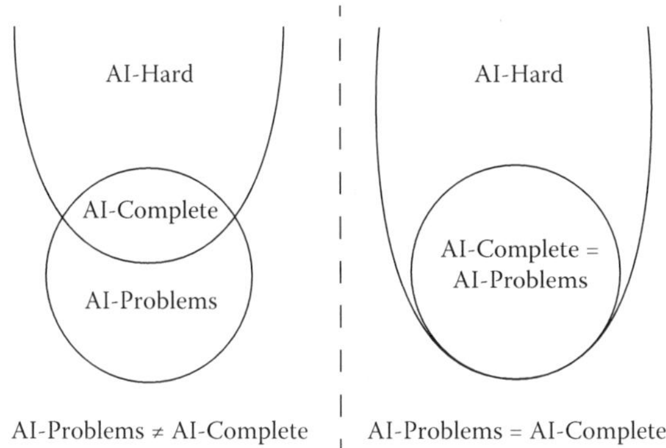

FIGURE 1.2 Relationship between AI complexity classes.

1.2.2 Turing Test as the First AI-Complete Problem

In this section, I show that a Turing test (TT; Turing 1950) problem is AI-Complete. First, I need to establish that a TT is indeed an AI problem (HO solvable). This trivially follows from the definition of the test itself. The test measures if a human-like performance is demonstrated by the test taker, and HOs are defined to produce human-level performance. While both *human* and *intelligence test* are intuitively understood terms, I have already shown that HOs could be expressed in strictly formal terms. The TT itself also could be formalized as an interactive proof (Shieber 2006, 2007; Bradford and Wollowski 1995).

The second requirement for a problem to be proven to be AI-Complete is that any other AI problem should be convertible into an instance of the problem under consideration in polynomial time via Turing reduction. Therefore, I need to show how any problem solvable by the Human function could be encoded as an instance of a TT. For any HO-solvable problem *h*, we have a string *input* that encodes the problem and a string *output* that encodes the solution. By taking the *input* as a question to be used in the TT and *output* as an answer to be expected while administering a TT, we can see how any HO-solvable problem could be reduced in polynomial time to an instance of a TT. Clearly, the described process is in polynomial time, and by similar algorithm, any AI problem could be reduced to TT. It is even theoretically possible to construct a complete TT that utilizes all other problems solvable by HO by generating one question from each such problem.

1.2.3 Reducing Other Problems to a TT

Having shown a first problem (TT) to be AI-Complete, the next step is to see if any other well-known AI problems are also AI-Complete. This is an effort similar to the work of Richard Karp, who showed some 21 problems were NP-Complete in his 1972 work and by doing so started a new field of computational complexity (Karp 1972). According to the *Encyclopedia of Artificial Intelligence* (Shapiro 1992), the following problems are all believed to be AI-Complete and so will constitute primary targets for our effort of proving formal AI-Completeness on them (Shapiro 1992, 54–57):

- **Natural Language Understanding:** "Encyclopedic knowledge is required to understand natural language. Therefore, a complete Natural Language system will also be a complete Intelligent system."
- **Problem Solving:** "Since any area investigated by AI researchers may be seen as consisting of problems to be solved, all of AI may be seen as involving Problem Solving and Search."
- **Knowledge Representation and Reasoning:** "The intended use is to use explicitly stored knowledge to produce additional explicit knowledge. This is what reasoning is. Together Knowledge representation and Reasoning can be seen to be both necessary and sufficient for producing general intelligence—it is another AI-complete area."
- **Vision or Image Understanding:** "If we take 'interpreting' broadly enough, it is clear that general intelligence may be needed to do this interpretation, and that correct interpretation implies general intelligence, so this is another AI-complete area."

Now that the TT has been proven to be AI-Complete, we have an additional way of showing other problems to be AI-Complete. We can either show that a problem is both in the set of AI problems and all other AI problems can be converted into it by some polynomial time algorithm or can reduce any instance of TT problem (or any other problem already proven to be AI-Complete) to an instance of a problem we are trying to show to be AI-Complete. This second approach seems to be particularly powerful. The general heuristic of my approach is to see if all information encoding the question that could be asked during administration of a TT could be encoded as an instance of a problem in question and likewise if any potential solution to that problem would constitute an answer to the relevant TT question. Under this heuristic, it is easy to see that, for example, chess is not AI-Complete as only limited information can be encoded as a starting position on a standard-size chessboard. Not surprisingly, chess has been one of the greatest successes of AI; currently, chess-playing programs dominate all human players, including world champions.

Question answering (QA) (Hirschman and Gaizauskas 2001; Salloum 2009) is a subproblem in natural language processing. Answering questions at a level of a human is something HOs are particularly good at based on their definition. Consequently, QA is an AI-Problem that is one of the two requirements for showing it to be AI-Complete. Having access to an oracle capable of solving QA allows us to solve TT via a simple reduction. For any statement *S* presented during administration of TT, transform said statement into a question for the QA oracle. The answers produced by the oracle can be used as replies in the TT, allowing the program to pass the TT. It is important to note that access to the QA oracle is sufficient to pass the TT only if questions are not restricted to stand-alone queries, but could contain information from previous questions. Otherwise, the problem is readily solvable even by today's machines, such as IBM's Watson, which showed a remarkable performance against human *Jeopardy* champions (Pepitone 2011).

Speech understanding (SU) (Anusuya and Katti 2009) is another subproblem in natural language processing. Understanding speech at a level of a human is something HOs are particularly good at based on their definition. Consequently, SU is an AI-Problem that is one of the two requirements for showing it to be AI-Complete. Having access to an oracle capable of solving SU allows us to solve QA via a simple reduction. We can reduce QA to SU by utilizing any text-to-speech software (Taylor and Black 1999; Chan 2003), which is both fast and accurate. This reduction effectively transforms written questions into the spoken ones, making it possible to solve every instance of QA by referring to the SU oracle.

1.2.4 Other Probably AI-Complete Problems

I hope that my work will challenge the AI community to prove other important problems as either belonging or not belonging to that class. Although the following problems have not been explicitly shown to be AI-Complete, they are strong candidates for such classification and are problems of great practical importance, making their classification a worthy endeavor. If a problem has been explicitly conjectured to be AI-Complete in a published paper, I include a source of such speculation: dreaming (Salloum 2009);

commonsense planning (Shahaf and Amir 2007); foreign policy (Mallery 1988); problem solving (Shapiro 1992); judging a TT (Shahaf and Amir 2007); commonsense knowledge (Andrich, Novosel, and Hrnkas 2009); SU (Shahaf and Amir 2007); knowledge representation and reasoning (Shapiro 1992); word sense disambiguation (Chen et al. 2009; Navigli and Velardi 2005); Machine Translation ("AI-Complete" 2011); ubiquitous computing (Leahu, Sengers, and Mateas 2008); change management for biomedical ontologies (Nejad 2010); NLU (Shapiro 1992); software brittleness ("AI-Complete" 2011); and vision or image understanding (Shapiro 1992).

1.3 FIRST AI-HARD PROBLEM: PROGRAMMING

I define the problem of programming as taking a natural language description of a program and producing a source code, which then is compiled on some readily available hardware/software to produce a computer program that satisfies all implicit and explicit requirements provided in the natural language description of the programming problem assignment. Simple examples of programming are typical assignments given to students in computer science classes, for example, "Write a program to play tic-tac-toe." Successful students write source code that, if correctly compiled, allows the grader to engage the computer in an instance of that game. Many requirements of such an assignment remain implicit, such as that response time of the computer should be less than a minute. Such implicit requirements are usually easily inferred by students who have access to culture-instilled common sense. As of this writing, no program is capable of solving programming outside strictly restricted domains.

Having access to an oracle capable of solving programming allows us to solve TT via a simple reduction. For any statement *S* presented during TT, transform said statement into a programming assignment of the form: "Write a program that would respond to *S* with a statement indistinguishable from a statement provided by an average human" (a full transcript of the TT may also be provided for disambiguation purposes). Applied to the set of all possible TT statements, this procedure clearly allows us to pass TT; however, programming itself is not in AI-Problems as there are many instances of programming that are not solvable by HOs. For example, "Write a program to pass a Turing test" is not known to be an AI-Problem under the proposed definition. Consequently, programming is an AI-Hard problem.

1.4 BEYOND AI-COMPLETENESS

The HO function presented in this chapter assumes that the human behind it has some assistance from the computer in order to process certain human unfriendly data formats. For example, a binary string representing a video is completely impossible for a human to interpret, but it could easily be played by a computer program in the intended format, making it possible for a human to solve a video understanding-related AI-Complete problem. It is obvious that a human provided with access to a computer (perhaps with Internet connection) is a more powerful intelligence compared to an unenhanced, in such a way, human. Consequently, it is important to limit help from a computer to a human worker "inside" a HO function to assistance in the domain of input/output conversion, but not beyond, as the resulting function would be both AI-Complete and "Computer Complete".

Figure 1.3 utilizes a Venn diagram to illustrate subdivisions of problem space produced by different types of intelligent computational devices. Region 1 represents what is known as a Universal Intelligence (Legg and Hutter 2007) or a Super Intelligence (Legg 2008; Yampolskiy 2011a, 2011b, 2012b)—a computational agent that outperforms all other intelligent

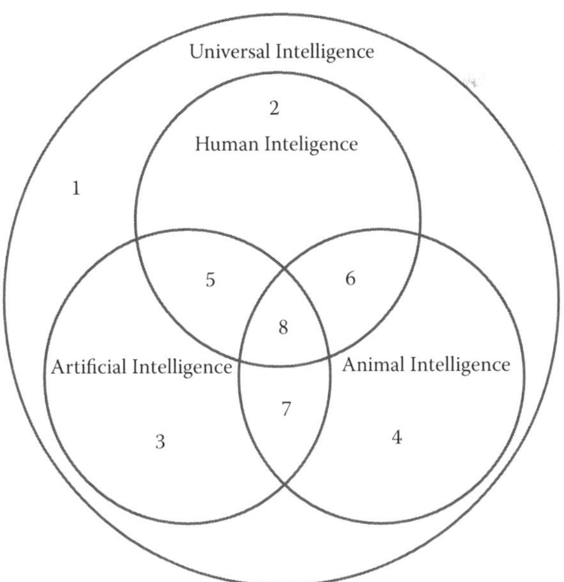

FIGURE 1.3 Venn diagram for four different types of intelligence.

agents over all possible environments. Region 2 is the standard unenhanced Human-level intelligence of the type capable of passing a TT, but at the same time incapable of computation involving large numbers or significant amount of memorization. Region 3 is what is currently possible to accomplish via state-of-the-art AI programs. Finally, Region 4 represents an abstract view of animal intelligence.

AI intelligence researchers strive to produce Universal Intelligence, and it is certainly likely to happen, given recent trends in both hardware and software developments and the theoretical underpinning of the Church/Turing Thesis (Turing 1936). It is also likely that if we are able to enhance human minds with additional memory and port those to a higher-speed hardware we will essentially obtain a Universal Intelligence (Sandberg and Boström 2008).

While the Universal Intelligence incorporates abilities of all the lower intelligences, it is interesting to observe that Human, AI and Animal intelligences have many interesting regions of intersection (Yampolskiy and Fox 2012). For example, animal minds are as good as human minds at visual understanding of natural scenes. Regions 5, 6, and 7 illustrate common problem spaces between two different types of intelligent agents. Region 8 represents common problem solving abilities of humans, computers and animals. Understanding such regions of commonality may help us to better separate the involved computational classes, which are represented by abilities of a specific computational agent minus the commonalities with a computational agent with which we are trying to draw a distinction. For example, CAPTCHA (Ahn et al. 2003) type tests rely on the inability of computers to perform certain pattern recognition tasks with the same level of accuracy as humans in order to separate AI agents from Human agents. Alternatively, a test could be devised to tell humans not armed with calculators from AIs by looking at the upper level of ability. Such a test should be easy to defeat once an effort is made to compile and formalize the limitations and biases of the human mind.

It is also interesting to consider the problem solving abilities of hybrid agents. I have already noted that a human being equipped with a computer is a lot more capable compared to an unaided person. Some research in Brain Computer Interfaces (Vidal 1973) provides a potential path for future developments in the area. Just as interestingly, combining pattern recognition abilities of animals with symbol processing abilities of AI could produce a computational agent with a large domain of human-like abilities (see work on RoboRats by Talwar et al. (2002) and on monkey controlled robots by Nicolelis et al. 2000). It is very likely that in the near future different types of intelligent agents will be combined to even greater extent. While such work is under way, I believe that it may be useful to introduce some additional terminology into the field of AI problem classification. For the complete space of problems I propose that the computational agents which are capable of solving a specific subset of such problems get to represent the set in question. Therefore, I propose additional terms: "Computer-Complete" and "Animals-Complete" to represent computational classes solvable by such agents. It is understood that just as humans differ in their abilities, so do animals and computers. Aggregation and averaging utilized in my Human function could be similarly applied to the definition of respective oracles. As research progresses, common names may be needed for different combinations of regions from Figure 1.3 illustrating such concepts as Human-AI hybrid or Animal-Robot hybrid.

Certain aspects of human cognition do not map well onto the space of problems which have seen a lot of success in the AI research field. Internal states of the human mind, such as consciousness (stream of), self-awareness, understanding, emotions (love, hate), feelings (pain, pleasure), etc., are not currently addressable by our methods. Our current state-of-the-art technologies are not sufficient to unambiguously measure or detect such internal states, and consequently even their existence is not universally accepted. Many scientists propose ignoring such internal states or claim they are nothing but a byproduct of flawed self-analysis. Such scientists want us to restrict science only to measurable behavioral actions; however, since all persons have access to internal states of at least one thinking machine, interest in trying to investigate internal states of the human mind is unlikely to vanish.

While I am able to present a formal theory of AI-Completeness based on the concept of HOs, the theory is not strong enough to address problems involving internal states of the mind. In fact, one of the fundamental arguments against our ability to implement understanding in a system that is based on symbol manipulation, Searle's Chinese Room thought experiment, itself relies on a generalized concept of a human as a part of a computational cycle. It seems that the current Turing/Von Neumann architecture is incapable of dealing with the set of problems which are related to internal states of human mind. Perhaps a new type of computational architecture capable of mimicking such internal states will be developed in the future. It is likely that it will be inspired by a better understanding of human biology and cognitive science. Research on creating Artificial Consciousness (AC) is attracting a lot of attention, at least in terms of number of AC papers published.

As a part of my ongoing effort to classify AI related problems, I propose a new category specifically devoted to problems of reproducing internal states of the human mind in artificial ways. I call this group of problems Consciousness-Complete or C-Complete for short. An oracle capable of solving C-Complete problems would be fundamentally different from the Oracle Machines proposed by Turing. C-Oracles would take input in the same way as their standard counterparts but would not produce any symbolic output. The result of their work would be a novel internal state of the oracle, which may become accessible to us if the new type of hardware discussed above is developed.

Just as SAT was shown to be the first NP-Complete problem and TT to be the first AI-Complete problem, I suspect that Consciousness will be shown to be the first C-Complete problem, with all other internal-state related problems being reducible to it. Which of the other internal state problems are also C-Complete is beyond the scope of this preliminary work. Even with no consciousness-capable hardware available at the moment of this writing, the theory of C-Completeness is still a useful tool, as it allows for formal classification of classical problems in the field of Artificial Intelligence into two very important categories: potentially *solvable* (with current technology) and unsolvable (with current technology). Since the only information available about HOs is their output and not internal states, they are fundamentally different from C-Oracles, creating two disjoint sets of problems.

The history of AI research is full of unwarranted claims of anticipated breakthroughs and, conversely, overestimations regarding the difficulty of some problems. Viewed through the prism of my AI-Complete/C-Complete theories, the history of AI starts to make sense. Solutions for problems that I classify as AI-Complete have been subject to continuous steady improvement, while those falling in the realm of C-Completeness have effectively seen zero progress (computer pain, Bishop 2009 and Dennett 1978; artificial consciousness, Searle 1980 and Dreyfus 1972; etc.). To proceed, science needs to better understand what the difference between a feeling and a thought is. Feeling pain and knowing about pain are certainly not the same internal states. I am hopeful that future research in this area will bring some long-awaited answers.

1.5 CONCLUSIONS

Progress in the field of artificial intelligence requires access to well-defined problems of measurable complexity. The theory of AI-Completeness aims to provide a base for such formalization. Showing certain problems to be

AI-Complete/-Hard is useful for developing novel ways of telling computers from humans. Also, any problem shown to be AI-Complete would be a great alternative way of testing an artificial intelligent agent to see if it attained human level intelligence (Shahaf and Amir 2007).

REFERENCES

Ahn, Luis von. June 2006. Games with a purpose. *IEEE Computer Magazine* 96–98.

Ahn, Luis von, Manuel Blum, Nick Hopper, and John Langford. 2003. CAPTCHA: Using Hard AI Problems for Security. Paper read at Eurocrypt. Advances in Cryptology — EUROCRYPT 2003. International Conference on the Theory and Applications of Cryptographic Techniques, Warsaw, Poland, May 4–8, 2003. Published in *Lecture Notes in Computer Science* 2656 (2003): 294–311.

AI-Complete. 2011. Accessed January 7. http://en.wikipedia.org/wiki/AI-complete.

Andrich, Christian, Leo Novosel, and Bojan Hrnkas. 2009. Common Sense Knowledge. Exercise Paper—Information Search and Retrieval. http://www.iicm.tu-graz.ac.at/cguetl/courses/isr/uearchive/uews2009/Ue06-CommonSenseKnowledge.pdf

Anusuya, M. A. and S. K. Katti. 2009. Speech recognition by machine: a review. *International Journal of Computer Science and Information Security (IJCSIS)* no. 6(3):181–205.

Bajaj, Vikas. April 25, 2010. Spammers pay others to answer security tests. *New York Times*.

Bergmair, Richard. December 2004. Natural Language Steganography and an "AI-Complete" Security Primitive. In 21st Chaos Communication Congress, Berlin.

Bishop, Mark. 2009. Why computers can't feel pain. *Minds and Machines* 19(4):507–516.

Bradford, Philip G. and Michael Wollowski. 1995. A formalization of the Turing Test. *SIGART Bulletin* 6(4):3–10.

Chan, Tsz-Yan. 2003. Using a text-to-speech synthesizer to generate a reverse Turing test. Paper presented at the 15th IEEE International Conference on Tools with Artificial Intelligence (ICTAI'03), Washington, DC, November 3–5.

Chen, Junpeng, Juan Liu, Wei Yu, and Peng Wu. November 30, 2009. Combining Lexical Stability and Improved Lexical Chain for Unsupervised Word Sense Disambiguation. Paper presented at the Second International Symposium on Knowledge Acquisition and Modeling (KAM '09), Wuhan, China.

Demasi, Pedro, Jayme L. Szwarcfiter, and Adriano J. O. Cruz. March 5–8, 2010. A Theoretical Framework to Formalize AGI-Hard Problems. Paper presented at the Third Conference on Artificial General Intelligence, Lugano, Switzerland.

Dennett, Daniel C. July 1978. Why you can't make a computer that feels pain. *Synthese* 38(3):415–456.

Dimmock, Nathan and Ian Maddison. December 2004. Peer-to-peer collaborative spam detection. *Crossroads* 11(2): 17–25.

Dreyfus, Hubert L. 1972. *What Computers Can't Do: A Critique of Artificial Reason*: New York: Harper & Row.

Gentry, Craig, Zulfikar Ramzan, and Stuart Stubblebine. June 5–8, 2005. Secure Distributed Human Computation. Paper presented at the 6th ACM Conference on Electronic Commerce, Vancouver, BC, Canada.

Hendler, James. September 2008. We've come a long way, maybe … . *IEEE Intelligent Systems* 23(5):2–3.

Hirschman, L., and R Gaizauskas. 2001. Natural language question answering. The view from here. *Natural Language Engineering* 7(4):275–300.

Horvitz, E. 2007. Reflections on challenges and promises of mixed-initiative interaction. *AI Magazine—Special Issue on Mixed-Initiative Assistants* 28(2): 11–18.

Horvitz, E. and T. Paek. 2007. Complementary computing: policies for transferring callers from dialog systems to human receptionists. *User Modeling and User Adapted Interaction* 17(1):159–182.

Ide, N. and J. Véronis. 1998. Introduction to the special issue on word sense disambiguation: the state of the art. *Computational Linguistics* 24(1):1–40.

Kapoor, A., D. Tan, P. Shenoy, and E. Horvitz. September 17–19, 2008. Complementary Computing for Visual Tasks: Meshing Computer Vision with Human Visual Processing. Paper presented at the IEEE International Conference on Automatic Face and Gesture Recognition, Amsterdam.

Karp, Richard M. 1972. Reducibility among combinatorial problems. In *Complexity of Computer Computations*, edited by R. E. Miller and J. W. Thatcher, 85–103. New York: Plenum.

Leahu, Lucian, Phoebe Sengers, and Michael Mateas. September 21–24, 2008. Interactionist AI and the Promise of ubicomp, or, How to Put Your Box in the World Without Putting the World in Your Box. Paper presented at the *Tenth International Conference on Ubiquitous Computing*. Seoul, South Korea.

Legg, Shane. June 2008. Machine Super Intelligence. PhD thesis, University of Lugano, Switzerland. http://www.vetta.org/documents/Machine_Super_Intelligence.pdf

Legg, Shane and Marcus Hutter. December 2007. Universal intelligence: a definition of machine intelligence. *Minds and Machines* 17(4):391–444.

Mallery, John C. 1988. Thinking about Foreign Policy: Finding an Appropriate Role for Artificial Intelligence Computers. Ph.D. dissertation, MIT Political Science Department, Cambridge, MA.

McIntire, John P., Paul R. Havig, and Lindsey K. McIntire. July 21–23, 2009. Ideas on Authenticating Humanness in Collaborative Systems Using AI-Hard Problems in Perception and Cognition. Paper presented at the IEEE National Aerospace and Electronics Conference (NAECON), Dayton, OH.

McIntire, John P., Lindsey K. McIntire, and Paul R. Havig. May 18–22, 2009. A Variety of Automated Turing tests for Network Security: Using AI-Hard Problems in Perception and Cognition to Ensure Secure Collaborations. Paper presented at the International Symposium on Collaborative Technologies and Systems (CTS '09), Baltimore.

Mert, Ezgi, and Cokhan Dalkilic. September 14–16, 2009. Word Sense Disambiguation for Turkish. Paper presented at the 24th International Symposium on Computer and Information Sciences (ISCIS 2009), Guzelyurt, Turkey.

Morgan, Nelson, D. Baron, S. Bhagat, H. Carvey, R. Dhillon, J. Edwards, D. Gelbart, A. Janin, A. Krupski, B. Peskin, T. Pfau, E. Shriberg, A. Stolcke, and C. Wooters. April 6–10, 2003. Meetings about Meetings: Research at ICSI on Speech in Multiparty Conversations. Paper presented at the IEEE International Conference on Acoustics, Speech, and Signal Processing (ICASSP '03), Hong Kong.

Mueller, Erik T. March 1987. Daydreaming and Computation. PhD dissertation, University of California, Los Angeles.

Navigli, Roberto, and Paola Velardi. July 2005. Structural semantic interconnections: a knowledge-based approach to word sense disambiguation. *IEEE Transactions on Pattern Analysis and Machine Intelligence* 27(7):1075–1086.

Nejad, Arash Shaban. April 2010. A Framework for Analyzing Changes in Health Care Lexicons and Nomenclatures. PhD dissertation, Concordia University, Montreal, QC, Canada.

Nicolelis, Miguel A. L., Johan Wessberg, Christopher R. Stambaugh, Jerald D. Kralik, Pamela D. Beck, Mark Laubach, John K. Chapin, and Jung Kim. 2000. Real-time prediction of hand trajectory by ensembles of cortical neurons in primates. *Nature* 408(6810):361.

Pepitone, Julianne. 2011. IBM's Jeopardy supercomputer beats humans in practice bout. *CNNMoney.* http://money.cnn.com/2011/01/13/technology/ibm_jeopardy_watson. Accessed January 13.

Phillips, P. Jonathon, and J. Ross Beveridge. September 28–30, 2009. An Introduction to Biometric-Completeness: The Equivalence of Matching and Quality. Paper presented at the IEEE 3rd International Conference on Biometrics: Theory, Applications, and Systems (BTAS '09), Washington, DC.

Raymond, Eric S. March 22, 1991. Jargon File Version 2.8.1. http://catb.org/esr/jargon/oldversions/jarg282.txt

Salloum, W. November 30, 2009. A Question Answering System Based on Conceptual Graph Formalism. Paper presented at the 2nd International Symposium on Knowledge Acquisition and Modeling (KAM 2009), Wuhan, China.

Sandberg, Anders, and Nick Boström. 2008. Whole Brain Emulation: A Roadmap. Technical Report 2008-3. Future of Humanity Institute, Oxford University. http://www.fhi.ox.ac.uk/Reports/2008-3.pdf

Searle, John. 1980. Minds, brains and programs. *Behavioral and Brain Sciences* 3(3):417–457.

Shahaf, Dafna, and Eyal Amir. March 26–28, 2007. Towards a Theory of AI Completeness. Paper presented at the 8th International Symposium on Logical Formalizations of Commonsense Reasoning (Commonsense 2007), Stanford University, Stanford, CA.

Shahaf, D., and E. Horvitz. July 2010. Generalized Task Markets for Human and Machine Computation. Paper presented at the Twenty-Fourth AAAI Conference on Artificial Intelligence, Atlanta, GA.

Shapiro, Stuart C. 1992. Artificial Intelligence. In *Encyclopedia of Artificial Intelligence*, edited by Stuart C. Shapiro, 54–57. New York: Wiley.

Shieber, Stuart M. July 16–20, 2006. Does the Turing Test Demonstrate Intelligence or Not? Paper presented at the Twenty-First National Conference on Artificial Intelligence (AAAI-06), Boston.

Shieber, Stuart M. December 2007. The Turing test as interactive proof. *Nous* 41(4):686–713.

Surowiecki, James. 2004. *The Wisdom of Crowds: Why the Many Are Smarter Than the Few and How Collective Wisdom Shapes Business, Economies, Societies and Nations*. New York: Little, Brown.

Takagi, H. 2001. Interactive evolutionary computation: fusion of the capacities of EC optimization and human evaluation. *Proceedings of the IEEE 89* 9:1275–1296.

Talwar, Sanjiv K., Shaohua Xu, Emerson S. Hawley, Shennan A. Weiss, Karen A. Moxon, and John K. Chapin. May 2, 2002. Behavioural neuroscience: rat navigation guided by remote control. *Nature* 417:37–38.

Taylor, P., and A. Black. 1999. Speech Synthesis by Phonological Structure Matching. Paper presented at Eurospeech99, Budapest, Hungary.

Turing, A. 1950. Computing machinery and intelligence. *Mind* 59(236):433–460.

Turing, Alan M. 1936. On computable numbers, with an application to the Entscheidungs problem. *Proceedings of the London Mathematical Society* 42:230–265.

Vaas, Lisa. December 1, 2007. Striptease used to recruit help in cracking sites. *PC Magazine*. http://www.pcmag.com/article2/0,2817,2210671,00.asp

Vidal, J. J. 1973. Toward direct brain-computer communication. *Annual Review of Biophysics and Bioengineering* 2:157–180.

Yampolskiy, R. V. 2011. AI-Complete CAPTCHAs as zero knowledge proofs of access to an artificially intelligent system. *ISRN Artificial Intelligence* 2012:271878.

Yampolskiy, Roman V. April 13, 2007a. Embedded CAPTCHA for Online Poker. Paper presented at the 20th Annual CSE Graduate Conference (Grad-Conf2007), Buffalo, NY.

Yampolskiy, Roman V. September 28, 2007b. Graphical CAPTCHA Embedded in Cards. Paper presented at the Western New York Image Processing Workshop (WNYIPW)—IEEE Signal Processing Society, Rochester, NY.

Yampolskiy, Roman V. October 3–4, 2011a. Artificial Intelligence Safety Engineering: Why Machine Ethics Is a Wrong Approach. Paper presented at Philosophy and Theory of Artificial Intelligence (PT-AI2011), Thessaloniki, Greece.

Yampolskiy, Roman V. October 3–4, 2011b. What to Do with the Singularity Paradox? Paper presented at Philosophy and Theory of Artificial Intelligence (PT-AI2011), Thessaloniki, Greece.

Yampolskiy, Roman V. April 21–22, 2012a. AI-Complete, AI-Hard, or AI-Easy—Classification of Problems in AI. Paper presented at the 23rd Midwest Artificial Intelligence and Cognitive Science Conference, Cincinnati, OH.

Yampolskiy, Roman V. 2012b. Leakproofing singularity—artificial intelligence confinement problem. *Journal of Consciousness Studies (JCS)* 19(1–2):194–214.

Yampolskiy, Roman V., and Joshua Fox. 2012. Artificial general intelligence and the human mental model. In *In the Singularity Hypothesis: A Scientific and Philosophical Assessment*, edited by Amnon Eden, Jim Moor, Johnny Soraker, and Eric Steinhart, 129–146. New York: Springer.

Yampolskiy, Roman V., and Venu Govindaraju. 2007. Embedded non-interactive continuous bot detection. *ACM Computers in Entertainment* 5(4):1–11.

CHAPTER 2

The Space of Mind Designs and the Human Mental Model

2.1 INTRODUCTION

In 1984, Aaron Sloman published "The Structure of the Space of Possible Minds," in which he described the task of providing an interdisciplinary description of that structure. He observed that "behaving systems" clearly comprise more than one sort of mind and suggested that virtual machines may be a good theoretical tool for analyzing mind designs. Sloman indicated that there are many discontinuities within the space of minds, meaning it is not a continuum or a dichotomy between things with minds and without minds (Sloman 1984). Sloman wanted to see two levels of exploration: *descriptive*, surveying things different minds can do, and *exploratory*, looking at how different virtual machines and their properties may explain results of the descriptive study (Sloman 1984). Instead of trying to divide the universe into minds and nonminds, he hoped to see examination of similarities and differences between systems. In this chapter, I make another step toward this important goal.

What is a mind? No universal definition exists. Solipsism notwithstanding, humans are said to have a mind. Higher-order animals are believed to have one as well, and maybe lower-level animals and plants, or even all life-forms. I believe that an artificially intelligent agent, such as a robot or a program running on a computer, will constitute a mind. Based on analysis of those examples, I can conclude that a mind is an instantiated intelligence with a knowledge base about its environment, and although intelligence itself is not an easy term to define, the work of Shane Legg provides a satisfactory, for my purposes, definition (Legg and Hutter 2007). In addition, some hold a point of view known as panpsychism, attributing mind-like properties to all matter. Without debating this possibility, I limit my analysis to those minds that can actively interact with their environment and other minds. Consequently, I do not devote any time to understanding what a rock is thinking.

If we accept materialism, we also have to accept that accurate software simulations of animal and human minds are possible. Those are known as uploads (Hanson 1994), and they belong to a class comprising computer programs no different from that to which designed or artificially evolved intelligent software agents would belong. Consequently, we can treat the space of all minds as the space of programs with the specific property of exhibiting intelligence if properly embodied. All programs could be represented as strings of binary numbers, implying that each mind can be represented by a unique number. Interestingly, Nick Bostrom, via some thought experiments, speculates that perhaps it is possible to instantiate a fractional number of mind, such as 0.3 mind, as opposed to only whole minds (Bostrom 2006). The embodiment requirement is necessary, because a string is not a mind, but could be easily satisfied by assuming that a universal Turing machine (UTM) is available to run any program we are contemplating for inclusion in the space of mind designs. An embodiment does not need to be physical as a mind could be embodied in a virtual environment represented by an avatar (Yampolskiy and Gavrilova 2012; Yampolskiy, Klare, and Jain 2012) and react to a simulated sensory environment like a "brain-in-a-vat" or a "boxed" artificial intelligence (AI) (Yampolskiy 2012b).

2.2 INFINITUDE OF MINDS

Two minds identical in terms of the initial design are typically considered to be different if they possess different information. For example, it is generally accepted that identical twins have distinct minds despite exactly the same blueprints for their construction. What makes them different is their individual experiences and knowledge obtained since inception. This implies that minds cannot be cloned because different copies would immediately after instantiation start accumulating different experiences and would be as different as twins.

If we accept that knowledge of a single unique fact distinguishes one mind from another, we can prove that the space of minds is infinite. Suppose we have a mind *M*, and it has a favorite number *N*. A new mind could be created by copying *M* and replacing its favorite number with a new favorite number $N + 1$. This process could be repeated infinitely, giving us an infinite set of unique minds. Given that a string of binary numbers represents an integer, we can deduce that the set of mind designs is an infinite and countable set because it is an infinite subset of integers. It is not the same as a set of integers because not all integers encode a mind.

Alternatively, instead of relying on an infinitude of knowledge bases to prove the infinitude of minds, we can rely on the infinitude of designs or embodiments. The infinitude of designs can be proven via inclusion of a time delay after every computational step. First, the mind would have a delay of 1 nanosecond, then a delay of 2 nanoseconds, and so on to infinity. This would result in an infinite set of different mind designs. Some will be very slow, others superfast, even if the underlying problem-solving abilities are comparable. In the same environment, faster minds would dominate slower minds proportionately to the difference in their speed. A similar proof with respect to the different embodiments could be presented by relying on an ever-increasing number of sensors or manipulators under control of a particular mind design.

Also, the same mind design in the same embodiment and with the same knowledge base may in fact effectively correspond to a number of different minds, depending on the operating conditions. For example, the same person will act differently if under the influence of an intoxicating substance, severe stress, pain, or sleep or food deprivation, or when experiencing a temporary psychological disorder. Such factors effectively change certain mind design attributes, temporarily producing a different mind.

2.3 SIZE, COMPLEXITY, AND PROPERTIES OF MINDS

Given that minds are countable, they could be arranged in an ordered list, for example, in order of numerical value of the representing string. This means that some mind will have the interesting property of being the smallest. If we accept that a UTM is a type of mind and denote by (m, n) the class of UTMs with m states and n symbols, the following UTMs have been discovered: (9, 3), (4, 6), (5, 5), and (2, 18). The (4, 6)-UTM uses only 22 instructions, and no less-complex standard machine has been found ("Universal Turing Machine" 2011). Alternatively, we may ask about the largest mind. Given that we have already shown that the set of minds is infinite, such an entity does not exist. However, if we take into account our embodiment requirement, the largest mind may in fact correspond to the design at the physical limits of computation (Lloyd 2000).

Another interesting property of minds is that they all can be generated by a simple deterministic algorithm, a variant of a Levin search (Levin 1973): Start with an integer (e.g., 42) and check to see if the number encodes a mind; if not, we discard the number. Otherwise, we add it to the set of mind designs and proceed to examine the next integer. Every mind will eventually appear on our list of minds after a predetermined number of steps. However, checking to see if something is in fact a mind is not a trivial procedure. Rice's theorem (Rice 1953) explicitly forbids determination of nontrivial properties of random programs. One way to overcome this limitation is to introduce an arbitrary time limit on the mind-or-not-mind determination function, effectively avoiding the underlying halting problem.

Analyzing our mind design generation algorithm, we may raise the question of a complexity measure for mind designs, not in terms of the abilities of the mind, but in terms of complexity of design representation. Our algorithm outputs minds in order of their increasing value, but this is not representative of the design complexity of the respective minds. Some minds may be represented by highly compressible numbers with a short representation such as 10^{13}, and others may comprise 10,000 completely random digits, for example, 735834895565117216037753562914 … (Yampolskiy 2013b). I suggest that a Kolmogorov complexity (KC) (Kolmogorov 1965) measure could be applied to strings representing mind designs. Consequently, some minds will be rated as "elegant" (i.e., having a compressed representation much shorter than the original string); others will be "efficient," representing the most efficient representation of that particular mind. Interesting elegant minds might be easier to discover than efficient minds, but unfortunately, KC is not generally computable.

In the context of complexity analysis of mind designs, we can ask a few interesting philosophical questions. For example, could two minds be added together (Sotala and Valpola 2012)? In other words, is it possible to combine two uploads or two artificially intelligent programs into a single, unified mind design? Could this process be reversed? Could a single mind be separated into multiple nonidentical entities, each in itself a mind? In addition, could one mind design be changed into another via a gradual process without destroying it? For example, could a computer virus (or even a real virus loaded with the DNA of another person) be a sufficient cause to alter a mind into a predictable type of other mind? Could specific properties be introduced into a mind given this virus-based approach? For example, could friendliness (Yudkowsky 2001) be added post-factum to an existing mind design?

Each mind design corresponds to an integer and so is finite, but because the number of minds is infinite, some have a much greater number of states compared to others. This property holds for all minds. Consequently, because a human mind has only a finite number of possible states, there are minds that can never be fully understood by a human mind, as such mind designs have a much greater number of states, making their understanding impossible, as can be demonstrated by the pigeonhole principle.

2.4 SPACE OF MIND DESIGNS

Overall, the set of human minds (about 7 billion of them currently available and about 100 billion that ever existed) is homogeneous in terms of both hardware (embodiment in a human body) and software (brain design and knowledge). In fact, the small differences between human minds are trivial in the context of the full infinite spectrum of possible mind designs. Human minds represent only a small constant-size subset of the great mind landscape. The same could be said about the sets of other earthly minds, such as dog minds, bug minds, male minds, or in general the set of all animal minds.

Given our algorithm for sequentially generating minds, one can see that a mind could never be completely destroyed, making minds theoretically immortal. A particular mind may not be embodied at a given time, but the idea of it is always present. In fact, it was present even before the material universe came into existence. So, given sufficient computational resources, any mind design could be regenerated, an idea commonly associated with the concept of reincarnation (Fredkin 1982). Also, the most powerful and most knowledgeable mind has always been associated with the idea of Deity or the Universal Mind.

Given my definition of mind, we can classify minds with respect to their design, knowledge base, or embodiment. First, the designs could be classified with respect to their origins: copied from an existing mind like an upload, evolved via artificial or natural evolution, or explicitly designed with a set of particular desirable properties. Another alternative is what is known as a Boltzmann brain—a complete mind embedded in a system that arises due to statistically rare random fluctuations in the particles comprising the universe, but is likely due to the vastness of the cosmos (De Simone et al. 2010).

Last, a possibility remains that some minds are physically or informationally recursively nested within other minds. With respect to the physical nesting, we can consider a type of mind suggested by Kelly (2007b), who talks about "a very slow invisible mind over large physical distances." It is possible that the physical universe as a whole or a significant part of it comprises such a megamind. This theory has been around for millennia and has recently received some indirect experimental support (Krioukov et al. 2012). In this case, all the other minds we can consider are nested within such a larger mind. With respect to the informational nesting, a powerful mind can generate a less-powerful mind as an idea. This obviously would take some precise thinking but should be possible for a sufficiently powerful artificially intelligent mind. Some scenarios describing informationally nested minds are analyzed in work on the AI confinement problem (Yampolskiy 2012b). Bostrom, using statistical reasoning, suggests that all observed minds, and the whole universe, are nested within a mind of a powerful computer (Bostrom 2003). Similarly, Lanza, using a completely different and somewhat controversial approach (biocentrism), argues that the universe is created by biological minds (Lanza 2007). It remains to be seen if given a particular mind, its origins can be deduced from some detailed analysis of the mind's design or actions.

Although minds designed by human engineers comprise only a tiny region in the map of mind designs, they probably occupy the best-explored part of the map. Numerous surveys of artificial minds, created by AI researchers in the last 50 years, have been produced (Miller 2012; Cattell and Parker 2012; de Garis et al. 2010; Goertzel et al. 2010; Vernon, Metta, and Sandini 2007). Such surveys typically attempt to analyze the state of the art in artificial cognitive systems and provide some internal classification of dozens of the reviewed systems with regard to their components and overall design. The main subcategories into which artificial minds designed by human engineers can be placed include neuron-level brain emulators (de Garis et al. 2010), biologically inspired cognitive architectures (Goertzel et al. 2010), physical symbol systems, emergent systems, and dynamical and enactive systems (Vernon, Metta, and Sandini 2007). Rehashing information about specific architectures presented in such surveys is beyond the scope of this book, but one can notice incredible richness and diversity of designs even in this tiny area of the overall map we are trying to envision. For those particularly interested in an overview of superintelligent minds, animal minds, and possible minds in addition to the surveys mentioned, "Artificial General Intelligence and the Human Mental Model" is highly recommended (Yampolskiy and Fox 2012).

For each mind subtype, there are numerous architectures, which to a certain degree depend on the computational resources available via a particular embodiment. For example, theoretically a mind working with infinite computational resources could trivially use brute force for any problem, always arriving at the optimal solution, regardless of its size. In practice, limitations of the physical world place constraints on available computational resources regardless of the embodiment type, making the brute force approach an infeasible solution for most real-world problems (Lloyd 2000). Minds working with limited computational resources have to rely on heuristic simplifications to arrive at "good enough" solutions (Yampolskiy, Ashby, and Hassan 2012; Ashby and Yampolskiy 2011; Hughes and Yampolskiy 2013; Port and Yampolskiy 2012).

Another subset of architectures consists of self-improving minds. Such minds are capable of examining their own design and finding improvements in their embodiment, algorithms, or knowledge bases that will allow the mind to more efficiently perform desired operations (Hall 2007b). It is likely that possible improvements would form a Bell curve with many initial opportunities for optimization toward higher efficiency and fewer such options remaining after every generation. Depending on the definitions used, one can argue that a recursively self-improving mind actually changes itself into a different mind, rather than remaining itself, which is particularly obvious after a sequence of such improvements. Taken to the extreme, this idea implies that a simple act of learning new information transforms you into a different mind, raising millennia-old questions about the nature of personal identity.

With respect to their knowledge bases, minds could be separated into those without an initial knowledge base, which are expected to acquire their knowledge from the environment; minds that are given a large set of universal knowledge from the inception; and those minds given specialized knowledge only in one or more domains. Whether the knowledge is stored in an efficient manner, compressed, classified, or censored is dependent on the architecture and is a potential subject of improvement by self-modifying minds.

One can also classify minds in terms of their abilities or intelligence. Of course, the problem of measuring intelligence is that no universal tests exist. Measures such as IQ tests and performance on specific tasks are not universally accepted and are always highly biased against nonhuman intelligences. Recently, some work has been done on streamlining intelligence measurements across different types of machine intelligence (Legg and Hutter 2007; Yonck 2012) and other "types" of intelligence (Herzing 2014), but the applicability of the results is still debated. In general, the notion of intelligence only makes sense in the context of problems to which said intelligence can be applied. In fact, this is exactly how IQ tests work—by presenting the subject with a number of problems and seeing how many the subject is able to solve in a given amount of time (computational resource).

A subfield of computer science known as computational complexity theory is devoted to studying and classifying various problems with respect to their difficulty and to the computational resources necessary to solve them. For every class of problems, complexity theory defines a class of machines capable of solving such problems. We can apply similar ideas to classifying minds; for example, all minds capable of efficiently (Yampolskiy 2013b) solving problems in the class P (polynomial) or a more difficult class of NP-complete problems (NP, nondeterministic polynomial time; Yampolskiy 2011b). Similarly, we can talk about minds with general intelligence belonging to the class of AI-Complete (Yampolskiy 2011a, 2012a, 2013c) minds, such as humans.

We can also look at the goals of different minds. It is possible to create a system that has no terminal goals, so such a mind is not motivated to accomplish things. Many minds are designed or trained for obtaining a particular high-level goal or a set of goals. We can envision a mind that has a randomly changing goal or a set of goals, as well as a mind that has many goals of different priority. Steve Omohundro used microeconomic theory to speculate about the driving forces in the behavior of superintelligent machines. He argues that intelligent machines will want to self-improve, be rational, preserve their utility functions, prevent counterfeit utility (Yampolskiy 2014), acquire and use resources efficiently, and protect themselves. He believes that the actions of machines will be governed by rational economic behavior (Omohundro 2007, 2008). Mark Waser suggests an additional "drive" to be included in the list of behaviors predicted to be exhibited by the machines (Waser 2010). Namely, he suggests that evolved desires for cooperation and being social are part of human ethics and are a great way of accomplishing goals, an idea also analyzed by Joshua Fox and Carl Shulman, but with contrary conclusions (Fox and Shulman 2010). Although it is commonly assumed that minds with high intelligence will converge on a common goal, Nick Bostrom, via his orthogonality thesis, has argued that a system can have any combination of intelligence and goals (Bostrom 2012).

Regardless of design, embodiment, or any other properties, all minds can be classified with respect to two fundamental but scientifically poorly defined properties: free will and consciousness. Both descriptors suffer from an ongoing debate regarding their actual existence or explanatory usefulness. This is primarily a result of the impossibility to design a definitive test to measure or even detect said properties, despite numerous attempts (Hales 2009; Aleksander and Dunmall 2003; Arrabales, Ledezma, and Sanchis 2008), or to show that theories associated with them are somehow falsifiable. Intuitively, we can speculate that consciousness, and maybe free will, are not binary properties but rather continuous and emergent abilities commensurate with a degree of general intelligence possessed by the system or some other property we shall term *mindness*. Free will can be said to correlate with a degree to which behavior of the system cannot be predicted (Aaronson 2013). This is particularly important in the design of artificially intelligent systems, for which inability to predict their future behavior is a highly undesirable property from the safety point of view (Yampolskiy 2013a, 2013d). Consciousness, on the other hand, seems to have no important impact on the behavior of the system, as can be seen from some thought experiments supposing existence of "consciousless" intelligent agents (Chalmers 1996). This may change if we are successful in designing a test, perhaps based on observer impact on quantum systems (Gao 2002), to detect and measure consciousness.

To be social, two minds need to be able to communicate, which might be difficult if the two minds do not share a common communication protocol, common culture, or even common environment. In other words, if they have no common grounding, they do not understand each other. We can say that two minds understand each other if, given the same set of inputs, they produce similar outputs. For example, in sequence prediction tasks (Legg 2006), two minds have an understanding if their predictions are the same regarding the future numbers of the sequence based on the same observed subsequence. We can say that a mind can understand another mind's function if it can predict the other's output with high accuracy. Interestingly, a perfect ability by two minds to predict each other would imply that they are identical and that they have no free will as defined previously.

2.5 A SURVEY OF TAXONOMIES

Yudkowsky describes the map of mind design space as follows: "In one corner, a tiny little circle contains all humans; within a larger tiny circle containing all biological life; and all the rest of the huge map is the space of minds-in-general. The entire map floats in a still vaster space, the space of optimization processes" (Yudkowsky 2008, 311). Figure 2.1 illustrates one possible mapping inspired by this description.

Similarly, Ivan Havel writes:

> All conceivable cases of intelligence (of people, machines, whatever) are represented by points in a certain abstract multi-dimensional "super space" that I will call the intelligence

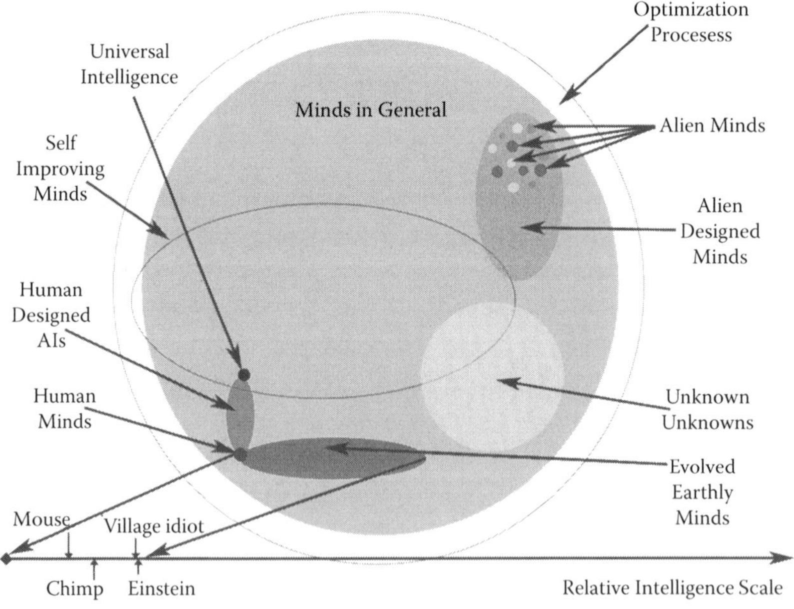

FIGURE 2.1 The universe of possible minds. (From Yudkowsky, Eliezer. May 13, 2006. Paper presented at the Singularity Summit at Stanford, Palo Alto, CA; and Yudkowsky, Eliezer. 2008. Artificial intelligence as a positive and negative factor in global risk. In *Global Catastrophic Risks*, edited by N. Bostrom and M. M. Cirkovic, 308–345. Oxford, UK: Oxford University Press.)

> space (shortly IS). Imagine that a specific coordinate axis in IS is assigned to any conceivable particular ability, whether human, machine, shared, or unknown (all axes having one common origin). If the ability is measurable the assigned axis is endowed with a corresponding scale. Hypothetically, we can also assign scalar axes to abilities, for which only relations like "weaker-stronger," "better-worse," "less-more" etc. are meaningful; finally, abilities that may be only present or absent may be assigned with "axes" of two (logical) values (yes-no). Let us assume that all coordinate axes are oriented in such a way that greater distance from the common origin always corresponds to larger extent, higher grade, or at least to the presence of the corresponding ability. The idea is that for each individual intelligence (i.e. the intelligence of a particular person, machine, network, etc.), as well as for each generic intelligence (of some group) there exists just one representing point in IS, whose coordinates determine the extent of involvement of particular abilities. (Havel 2013, 13)

If the universe (or multiverse) is infinite, as our current physics theories indicate, then all possible minds in all possible states are instantiated somewhere (Bostrom 2006).

Ben Goertzel proposes the following classification of kinds of minds, mostly centered on the concept of embodiment (Geortzel 2006):

- **Singly embodied:** controls a single physical or simulated system
- **Multiply embodied:** controls a number of disconnected physical or simulated systems
- **Flexibly embodied:** controls a changing number of physical or simulated systems
- **Nonembodied:** resides in a physical substrate but does not utilize the body in a traditional way
- **Body centered:** consists of patterns emergent between the physical system and the environment
- **Mindplex:** consists of a set of collaborating units, each of which is itself a mind (Goertzel 2003)

- **Quantum:** is an embodiment based on properties of quantum physics
- **Classical:** is an embodiment based on properties of classical physics

J. Storrs Hall (2007a), in his "Kinds of Minds," suggests that different stages to which a developing AI might belong can be classified relative to its humanlike abilities. His classification encompasses the following:

- **Hypohuman:** infrahuman, less-than-human capacity
- **Diahuman:** human-level capacities in some areas but still no general intelligence
- **Parahuman:** similar but not identical to humans, as for example, augmented humans
- **Allohuman:** as capable as humans, but in different areas
- **Epihuman:** slightly beyond the human level
- **Hyperhuman:** much more powerful than human; superintelligent (Hall 2007a; Yampolskiy and Fox 2012)

Patrick Roberts, in his book *Mind Making*, presents his ideas for a "taxonomy of minds"; I leave it to the reader to judge the usefulness of his classification (Roberts 2009):

- **Choose means:** Does it have redundant means to the same ends? How well does it move between them?
- **Mutate:** Can a mind naturally gain and lose new ideas in its lifetime?
- **Doubt:** Is it eventually free to lose some or all beliefs? Or, is it wired to obey the implications of every sensation?
- **Sense itself:** Does a mind have the senses to see the physical conditions of that mind?
- **Preserve itself:** Does a mind also have the means to preserve or reproduce itself?
- **Sense minds:** Does a mind understand a mind, at least of lower classes, and how well does it apply that to itself, to others?

- **Sense kin:** Can it recognize the redundant minds, or at least the bodies of minds, with which it was designed to cooperate?
- **Learn:** Does the mind's behavior change from experience? Does it learn associations?
- **Feel:** We imagine that an equally intelligent machine would lack our conscious experience.
- **Communicate:** Can it share beliefs with other minds?

Kevin Kelly has also proposed a taxonomy of minds that in his implementation is really just a list of different minds, some of which have not appeared in other taxonomies (Kelly 2007b): mind with operational access to its source code; general intelligence without self-awareness; self-awareness without general intelligence; superlogic machine without emotion; mind capable of imagining greater mind; self-aware mind incapable of creating a greater mind; mind capable of creating a greater mind, which creates a greater mind; very slow, distributed mind over large physical distance; mind capable of cloning itself and remaining in unity with clones; global mind, which is a large supercritical mind of subcritical brains; and anticipator, mind specializing in scenario and prediction making (Kelly 2007b).

Elsewhere, Kelly provides much relevant analysis of the landscape of minds and writes about "Inevitable Minds" (Kelly 2009), "The Landscape of Possible Intelligences" (Kelly 2008a), "What Comes After Minds?" (Kelly 2008b), and "The Evolutionary Mind of God" (Kelly 2007a).

Aaron Sloman, in "The Structure of the Space of Possible Minds," using his virtual machine model, proposes a division of the space of possible minds with respect to the following properties (Sloman 1984):

- Quantitative versus structural
- Continuous versus discrete
- Complexity of stored instructions
- Serial versus parallel
- Distributed versus fundamentally parallel
- Connected to external environment versus not connected
- Moving versus stationary

- Capable of modeling others versus not capable
- Capable of logical inference versus not capable
- Fixed versus reprogrammable
- Goal consistency versus goal selection
- Metamotives versus motives
- Able to delay goals versus immediate goal following
- Statics plan versus dynamic plan
- Self-aware versus not self-aware

2.6 MIND CLONING AND EQUIVALENCE TESTING ACROSS SUBSTRATES

The possibility of uploads rests on the ideas of computationalism (Putnam 1980), specifically substrate independence and equivalence, meaning that the same mind can be instantiated in different substrates and move freely between them. If your mind is cloned and if a copy is instantiated in a different substrate from the original one (or on the same substrate), how can it be verified that the copy is indeed an identical mind? Can it be done at least immediately after cloning and before the mind-clone learns any new information? For that purpose, I propose a variant of a Turing test, which also relies on interactive text-only communication to ascertain the quality of the copied mind. The text-only interface is important not to prejudice the examiner against any unusual substrates on which the copied mind might be running. The test proceeds by having the examiner (original mind) ask questions of the copy (cloned mind), questions that supposedly only the original mind would know answers to (testing should be done in a way that preserves privacy). Good questions would relate to personal preferences, secrets (passwords, etc.), as well as recent dreams. Such a test could also indirectly test for consciousness via similarity of subjective qualia. Only a perfect copy should be able to answer all such questions in the same way as the original mind. Another variant of the same test may have a third party test the original and cloned mind by seeing if they always provide the same answer to any question. One needs to be careful in such questioning not to give undue weight to questions related to the mind's substrate as that may lead to different answers. For example, asking a human if he or she is hungry may produce an answer different from the one that would be given by a nonbiological robot.

2.7 CONCLUSIONS

Science periodically experiences a discovery of a whole new area of investigation. For example, observations made by Galileo Galilei led to the birth of observational astronomy (Galilei 1953), also known as the study of our universe; Watson and Crick's discovery of the structure of DNA led to the birth of the field of genetics (Watson and Crick 1953), which studies the universe of blueprints for organisms; and Stephen Wolfram's work with cellular automata has resulted in "a new kind of science" (Wolfram 2002) that investigates the universe of computational processes. I believe that we are about to discover yet another universe: the universe of minds.

As our understanding of the human brain improves, thanks to numerous projects aimed at simulating or reverse engineering a human brain, we will no doubt realize that human intelligence is just a single point in the vast universe of potential intelligent agents comprising a new area of study. The new field, which I would like to term *intellectology*, will study and classify design space of intelligent agents, work on establishing limits to intelligence (minimum sufficient for general intelligence and maximum subject to physical limits), contribute to consistent measurement of intelligence across intelligent agents, look at recursive self-improving systems, design new intelligences (making AI a subfield of intellectology), and evaluate the capacity for understanding higher-level intelligences by lower-level ones. At the more theoretical level, the field will look at the distribution of minds on the number line and in the mind design space, as well as attractors in the mind design space. It will consider how evolution, drives, and design choices have an impact on the density of minds in the space of possibilities. The field will not be subject to the current limitations brought on by the human-centric view of intelligence and will open our understanding to seeing intelligence as a fundamental computational resource, such as space or time. Finally, I believe intellectology will highlight the inhumanity of most possible minds and the dangers associated with such minds being placed in charge of humanity.

REFERENCES

Aaronson, Scott. 2013. The Ghost in the Quantum Turing Machine. arXiv preprint arXiv:1306.0159.

Aleksander, Igor and Barry Dunmall. 2003. Axioms and tests for the presence of minimal consciousness in agents I: preamble. *Journal of Consciousness Studies* 10(4–5):4–5.

Arrabales, Raúl, Agapito Ledezma, and Araceli Sanchis. 2008. ConsScale: a plausible test for machine consciousness? *Proceedings of the Nokia Workshop on Machine Consciousness—13th Finnish Artificial Intelligence Conference (STeP 2008), Helsinki, Finland*, pp. 49–57.

Ashby, Leif H. and Roman V. Yampolskiy. July 27–30, 2011. Genetic algorithm and wisdom of artificial crowds algorithm applied to Light up. Paper presented at the 2011 16th International Conference on Computer Games (CGAMES), Louisville, KY.

Bostrom, Nick. 2003. Are you living in a computer simulation? *Philosophical Quarterly* 53(211):243–255.

Bostrom, Nick. 2006. Quantity of experience: brain-duplication and degrees of consciousness. *Minds and Machines* 16(2):185–200.

Bostrom, Nick. 2012. The superintelligent will: motivation and instrumental rationality in advanced artificial agents. *Minds and Machines* 22(2):71–85.

Cattell, Rick and Alice Parker. 2012. Challenges for brain emulation: why is it so difficult? *Natural Intelligence* 1(3):17–31.

Chalmers, David J. 1996. *The Conscious Mind: In Search of a Fundamental Theory*. Oxford, UK: Oxford University Press.

de Garis, Hugo, Chen Shuo, Ben Goertzel, and Lian Ruiting. 2010. A world survey of artificial brain projects. Part I: large-scale brain simulations. *Neurocomputing* 74(1–3):3–29. doi:http://dx.doi.org/10.1016/j.neucom.2010.08.004

De Simone, Andrea, Alan H. Guth, Andrei Linde, Mahdiyar Noorbala, Michael P. Salem, and Alexander Vilenkin. 2010. Boltzmann brains and the scale-factor cutoff measure of the multiverse. *Physical Review D* 82(6):063520.

Fox, Joshua and Carl Shulman. October 4–6, 2010. Superintelligence Does Not Imply Benevolence. Paper presented at the Eighth European Conference on Computing and Philosophy, Munich, Germany.

Fredkin, Edward. 1982. On the Soul. Unpublished manuscript.

Galilei, Galileo. 1953. *Dialogue Concerning the Two Chief World Systems: Ptolemaic and Copernican*. Oakland: University of California Press.

Gao, Shan. 2002. A quantum method to test the existence of consciousness. *The Noetic Journal* 3(3):27–31.

Goertzel, Ben. 2003. Mindplexes: the potential emergence of multiple levels of focused consciousness in communities of AI's and humans. *Dynamical Psychology*. http://www.goertzel.org/dynapsyc/2003/mindplex.htm

Goertzel, Ben. 2006. Kinds of minds. In *The Hidden Pattern: A Patternist Philosophy of Mind*, 17–25. Boca Raton, FL: BrownWalker Press.

Goertzel, Ben, Ruiting Lian, Itamar Arel, Hugo de Garis, and Shuo Chen. 2010. A world survey of artificial brain projects. Part II: biologically inspired cognitive architectures. *Neurocomputing* 74(1–3):30–49. doi:10.1016/j.neucom.2010.08.012

Hales, Colin. 2009. An empirical framework for objective testing for P-consciousness in an artificial agent. *Open Artificial Intelligence Journal* 3:1–15.

Hall, J. Storrs. 2007a. Kinds of minds. In *Beyond AI: Creating the Conscience of the Machine*, 241–248. Amherst, NY: Prometheus Books.

Hall, J. Storrs. October 2007b. Self-improving AI: an analysis. *Minds and Machines* 17(3):249–259.

Hanson, Robin. 1994. If uploads come first. *Extropy* 6(2): 10–15.

Havel, Ivan M. 2013. On the way to intelligence singularity. In *Beyond Artificial Intelligence*, edited by Jozef Kelemen, Jan Romportl, and Eva Zackova, 3–26. Berlin: Springer.

Herzing, Denise L. 2014. Profiling nonhuman intelligence: an exercise in developing unbiased tools for describing other "types" of intelligence on earth. *Acta Astronautica* 94(2):676–680. doi:http://dx.doi.org/10.1016/j.actaastro.2013.08.007

Hughes, Ryan, and Roman V. Yampolskiy. 2013. Solving sudoku puzzles with wisdom of artificial crowds. *International Journal of Intelligent Games and Simulation* 7(1):6.

Kelly, Kevin. 2007a. The Evolutionary Mind of God. http://kk.org/thetechnium/archives/2007/02/the_evolutionar.php

Kelly, Kevin. 2007b. A Taxonomy of Minds. http://kk.org/thetechnium/archives/2007/02/a_taxonomy_of_m.php

Kelly, Kevin. 2008a. The Landscape of Possible Intelligences. http://kk.org/thetechnium/archives/2008/09/the_landscape_o.php

Kelly, Kevin. 2008b. What Comes After Minds? http://kk.org/thetechnium/archives/2008/12/what_comes_afte.php

Kelly, Kevin. 2009. Inevitable Minds. http://kk.org/thetechnium/archives/2009/04/inevitable_mind.php

Kolmogorov, A. N. 1965. Three approaches to the quantitative definition of information. *Problems of Information Transmission* 1(1):1–7.

Krioukov, Dmitri, Maksim Kitsak, Robert S. Sinkovits, David Rideout, David Meyer, and Marián Boguñá. 2012. Network cosmology. *Science Reports* 2. doi:http://www.nature.com/srep/2012/121113/srep00793/abs/srep00793.html#supplementary-information

Lanza, Robert. 2007. A new theory of the universe. *American Scholar* 76(2):18.

Legg, Shane. 2006. Is There an Elegant Universal Theory of Prediction? Paper presented at Algorithmic Learning Theory. 17th International Conference on Algorithmic Learning Theory, Barcelona, Spain, October 7–10, 2006.

Legg, Shane and Marcus Hutter. December 2007. Universal intelligence: a definition of machine intelligence. *Minds and Machines* 17(4):391–444.

Levin, Leonid. 1973. Universal search problems. *Problems of Information Transmission* 9(3):265–266.

Lloyd, Seth. 2000. Ultimate physical limits to computation. *Nature* 406:1047–1054.

Miller, M. S. P. July 4–6, 2012. Patterns for Cognitive Systems. Paper presented at the 2012 Sixth International Conference on Complex, Intelligent and Software Intensive Systems (CISIS), Palermo, Italy.

Omohundro, Stephen M. September 8–9, 2007. The Nature of Self-Improving Artificial Intelligence. Paper presented at the Singularity Summit, San Francisco.

Omohundro, Stephen M. February 2008. The Basic AI Drives. In *Proceedings of the First AGI Conference, Volume 171, Frontiers in Artificial Intelligence and Applications*, edited by P. Wang, B. Goertzel, and S. Franklin, 483–492. Amsterdam: IOS Press.

Port, Aaron C. and Roman V. Yampolskiy. July 30–August 1, 2012. Using a GA and wisdom of artificial crowds to solve solitaire battleship puzzles. Paper presented at the 2012 17th International Conference on Computer Games (CGAMES), Louisville, KY.

Putnam, Hilary. 1980. Brains and behavior. *Readings in Philosophy of Psychology* 1:24–36.

Rice, Henry Gordon. 1953. Classes of recursively enumerable sets and their decision problems. *Transactions of the American Mathematical Society* 74(2):358–366.

Roberts, Patrick. 2009. *Mind Making: The Shared Laws of Natural and Artificial.* North Charleston, SC: CreateSpace.

Sloman, Aaron. 1984. The structure of the space of possible minds. In *The Mind and the Machine: Philosophical Aspects of Artificial Intelligence*, 35–42. Chichester, UK: Ellis Horwood.

Sotala, Kaj and Harri Valpola. 2012. Coalescing minds: brain uploading-related group mind scenarios. *International Journal of Machine Consciousness* 4(1):293–312. doi:10.1142/S1793843012400173

Universal Turing Machine. 2011. Accessed April 14. http://en.wikipedia.org/wiki/Universal_Turing_machine

Vernon, D., G. Metta, and G. Sandini. 2007. A survey of artificial cognitive systems: implications for the autonomous development of mental capabilities in computational agents. *IEEE Transactions on Evolutionary Computation* 11(2):151–180. doi:10.1109/tevc.2006.890274

Waser, Mark R. March 5–8, 2010. Designing a Safe Motivational System for Intelligent Machines. Paper presented at the Third Conference on Artificial General Intelligence, Lugano, Switzerland.

Watson, James D. and Francis H. C. Crick. 1953. Molecular structure of nucleic acids. *Nature* 171(4356):737–738.

Wolfram, Stephen. May 14, 2002. *A New Kind of Science.* Oxfordshire, UK: Wolfram Media.

Yampolskiy, R. V. 2011a. AI-Complete CAPTCHAs as zero knowledge proofs of access to an artificially intelligent system. *ISRN Artificial Intelligence* no. 271878.

Yampolskiy, Roman V. 2011b. Construction of an NP problem with an exponential lower bound. Arxiv preprint arXiv:1111.0305

Yampolskiy, Roman V. April 21–22, 2012a. AI-Complete, AI-Hard, or AI-Easy—Classification of Problems in AI. Paper presented at the 23rd Midwest Artificial Intelligence and Cognitive Science Conference, Cincinnati, OH.

Yampolskiy, Roman V. 2012b. Leakproofing singularity—artificial intelligence confinement problem. *Journal of Consciousness Studies (JCS)* 19(1–2):194–214.

Yampolskiy, Roman V. 2013a. Artificial intelligence safety engineering: why machine ethics is a wrong approach. In *Philosophy and Theory of Artificial Intelligence*, 389–396. Berlin: Springer.

Yampolskiy, Roman V. 2013b. Efficiency theory: a unifying theory for information, computation and intelligence. *Journal of Discrete Mathematical Sciences and Cryptography* 16(4–5):259–277.

Yampolskiy, Roman V. 2013c. Turing test as a defining feature of AI-Completeness. In *Artificial Intelligence, Evolutionary Computation and Metaheuristics—In the Footsteps of Alan Turing,* edited by Xin-She Yang, 3–17. Berlin: Springer.

Yampolskiy, Roman V. 2013d. What to do with the singularity paradox? In *Philosophy and Theory of Artificial Intelligence*, 397–413. Berlin: Springer.

Yampolskiy, Roman V. 2014. Utility function security in artificially intelligent agents. *Journal of Experimental and Theoretical Artificial Intelligence (JETAI)* 1–17.

Yampolskiy, Roman V., Leif Ashby, and Lucas Hassan. 2012. Wisdom of artificial crowds—a metaheuristic algorithm for optimization. *Journal of Intelligent Learning Systems and Applications* 4(2):98–107.

Yampolskiy, Roman V., and Joshua Fox. 2012. Artificial general intelligence and the human mental model. In *Singularity Hypotheses*, 129–145. Berlin: Springer.

Yampolskiy, Roman and Marina Gavrilova. 2012. Artimetrics: biometrics for artificial entities. *IEEE Robotics and Automation Magazine (RAM)* 19(4):48–58.

Yampolskiy, Roman V., Brendan Klare, and Anil K. Jain. December 12–15, 2012. Face Recognition in the Virtual World: Recognizing Avatar Faces. Paper presented at the 2012 11th International Conference on Machine Learning and Applications (ICMLA), Boca Raton, FL.

Yonck, Richard. 2012. Toward a standard metric of machine intelligence. *World Future Review* 4(2):61–70.

Yudkowsky, Eliezer S. 2001. Creating Friendly AI—The Analysis and Design of Benevolent Goal Architectures. http://singinst.org/upload/CFAI.html

Yudkowsky, Eliezer. May 13, 2006. The Human Importance of the Intelligence Explosion. Paper presented at the Singularity Summit at Stanford, Palo Alto, CA.

Yudkowsky, Eliezer. 2008. Artificial intelligence as a positive and negative factor in global risk. In *Global Catastrophic Risks*, edited by N. Bostrom and M. M. Cirkovic, 308–345. Oxford, UK: Oxford University Press.

CHAPTER 3

How to Prove You Invented Superintelligence So No One Else Can Steal It

3.1 INTRODUCTION AND MOTIVATION

Experts predict that in the next 10 to 100 years scientists will succeed in creating human-level artificial general intelligence (AGI) (Yudkowsky 2008; Bostrom 2006; Hibbard 2005; Chalmers 2010; Hall 2000). Although it is most likely that AGI will be created by a government agency (Shulman 2009) such as the Defense Advanced Research Projects Agency (DARPA) or a large corporation such as Google, the possibility remains that it will be done by a single inventor or a small team of "garage inventors." The history of computer science is the history of such inventors. Steve Jobs and Steve Wozniak (Apple), Bill Gates (Microsoft), Mark Zuckerberg (Facebook), Larry Page and Sergey Brin (Google), to name just a few, all revolutionized the state of technology while they were independent inventors.

What is an inventor to do after successful construction of an artificially intelligent system? Going public with such an invention may be dangerous as numerous powerful entities will try to steal the invention. Worse yet, they will also likely try to reduce the inventor's freedom and safety either to prevent leaking of information or to secure necessary assistance in understanding the invention. Potential nemeses include security agencies, government representatives, the military complex, multinational corporations, competing scientists, foreign governments, and potentially anyone who understands the value of such an invention.

It has been said that a true artificial intelligence (AI) is the last invention we will ever have to make (Good 1966): It will make all other inventions for us. The monetary value of a true AI system is hard to overestimate, but it is well known that billions already have been spent on research by governments and industry (Russell and Norvig 2003). Its potential for the military complex is unprecedented in terms of both smart weapons and human-free combat (Sparrow 2007). Even if the initial system has only human-level intelligence, such a machine, among other things, would be capable of designing the next generation of even smarter intelligent machines; it is generally assumed that an intelligence explosion will take place shortly after such a technological self-improvement cycle begins leading to creation of superintelligence. Possession of such a system would clearly put the inventor of the system in danger (Good 1966).

In this chapter, I address the question of safeguarding a true AI, a discovery that could without hesitation be said to be worth trillions of dollars. Without going into details, I assume that the inventor, through code obfuscation, encryption, anonimization, and location obscurity, is able to prevent others from directly accessing the system but still wishes to prove that it was constructed. For this purpose, I propose a novel method based on the combination of zero-knowledge proofs (ZKPs) and provably AI-Complete CAPTCHA (Completely Automated Public Turing Test to Tell Computers and Humans Apart) problems to show that a superintelligent system has been constructed without having to reveal the design of the system.

Alternatively, my method could be used to convince a group of skeptics that in fact a true AI system has been invented without having to resort to time-consuming individual demonstrations. This would be useful if the inventor faces a skeptical reception from the general public and scientific community. In the past, exaggerated claims have been made (Russell and Norvig 2003) about some AI systems, so a skeptical reception would not be that surprising. The following sections provide an overview of ZKPs, CAPTCHAs, and the concept of AI-Completeness, all of which are necessary to understand the proposed method.

3.2 ZERO KNOWLEDGE PROOF

Simply stated, a ZKP is an interactive probabilistic protocol between two parties that gives, with a high degree of certainty, evidence that a theorem is true and that the prover knows a proof while providing not a single bit of information about the said proof to the verifier (Blum 1986). The ZKP works by breaking up the proof into several pieces in such a way that (Blum 1986)

1. The verifier can tell whether any given piece of the proof is properly constructed.
2. The combination of all the pieces constitutes a valid proof.
3. Revealing any single piece of the proof does not reveal any information about the proof.

To begin, the prover hides each piece of the proof by applying a one-way function to it. After that, the verifier is allowed to request a decryption of any single piece of the proof. Because the verifier can select a specific piece at random, seeing that it is properly constructed provides probabilistic evidence that all pieces of the proof are properly constructed and so is the proof as the whole (Blum 1986).

3.3 CAPTCHA*

With the steady increase in popularity of services offered via the Internet, the problem of securing such services from automated attacks became apparent (Yampolskiy and Govindaraju 2007). To protect limited computational resources against utilization by the growing number of human-impersonating artificially intelligent systems, a methodology was necessary to discriminate between such systems and people (Pope and Kaur 2005). In 1950, Alan Turing published his best-known paper, "Computing Machinery and Intelligence," in which he proposed evaluating the abilities of an artificially intelligent machine based on how closely it can mimic human behavior. The test, which is now commonly known as the Turing test, is structured as a conversation and can be used to evaluate multiple behavioral parameters, such as an agent's knowledge, skills, preferences, and strategies (French 2000). In essence, it is the ultimate

* Roman V. Yampolskiy and Venu Govindaraju, *ACM Computers in Entertainment* 5(4):1–11, 2007. http://doi.acm.org/10.1145/1324198.1324205 Copyright 2007 ACM, Inc. Reprinted by permission.

multimodal behavioral biometric, which was postulated to make it possible to detect differences between a human and a machine (Yampolskiy and Govindaraju 2007).

The theoretical platform for an automated Turing test (ATT) was developed by Moni Naor in 1996. The following properties were listed as desirable for the class of problems that can serve as an ATT:

- Many instances of a problem can be automatically generated together with their solutions.
- Humans can solve any instance of a problem quickly and with a low error rate. The answer should be easy to provide either by a menu selection or via typing a few characters.
- The best known AI programs for solving such problems fail a significant percentage of times, despite the full disclosure of how the test problem is generated.
- The test problem specification needs to be concise in terms of description and area used to present the test to the user.

Since the initial paper by Naor, a great deal of research has been performed in the area, with different researchers frequently inventing new names for the same concept of human/machine disambiguation (Baird and Popat 2002; Sampson 2006). In addition to the ATT, the developed procedures are known under such names as (Yampolskiy and Govindaraju 2007) reversed Turing test (RTT; Coates, Baird, and Fateman 2001); human interactive proof (HIP; Chellapilla et al. 2005a); mandatory human participation (MHP; Xu et al. 2003); or CAPTCHA (Ahn, Blum, and Langford 2004; Ahn 2004).

As ongoing developments in AI research allow some tests to be broken (Chellapilla and Simard 2004; Mori and Malik 2003; Aboufadel, Olsen, and Windle 2005; Moy et al. 2004), research continues on developing ways of telling machines and humans apart that are more secure and user friendly (Rui et al. 2005; Chellapilla et al. 2005b, 2005c; Wang, Baird, and Bentley 2006; May 2005; Lopresti 2005). Such tests are always based on an as-yet-unsolved problem in AI (Ahn et al. 2003). Frequent examples include pattern recognition, in particular character recognition (Bentley and Mallows 2006; Baird and Riopka 2005; Baird, Moll, and Wang 2005a, 2005b; Chew and Baird 2003; Simard et al. 2003; Liao and Chang 2004) or image recognition (Chew and Tygar 2004; Liao 2006; Dailey and Namprempre 2004). A number of CAPTCHAs are based on recognition of different biometrics, such as faces (Misra and Gaj 2006; Rui and Liu 2003a, 2003b), voice (Kochanski, Lopresti, and Shih 2002; Chan 2003), or handwriting (Rusu and Govindaraju 2004, 2005). In addition, experimentation has occurred with the following types of tests (Hall 2006; Yampolskiy and Govindaraju 2007):

- **Reading:** Display of a password as a cluttered image
- **Shape:** Identification of complex shapes
- **Spatial:** Rendering of a text image from a three-dimensional (3-D) model
- **Quiz:** Display of a visual or audio puzzle or trivia question
- **Match:** Common theme identification for a set of related images
- **Virtual Reality:** Navigation in a 3-D world
- **Natural:** Use of media files collected from the real world, particularly the web
- **Implicit:** Incorporation of the test into the web page navigation system (Baird and Bentley 2005)

3.4 AI-COMPLETENESS*

A somewhat general definition of the term *AI-Complete* included in the 1991 Jargon File (Raymond 1991) states the following:

> AI-complete: [MIT, Stanford, by analogy with "NP-complete"] adj. Used to describe problems or subproblems in AI, to indicate that the solution presupposes a solution to the "strong AI problem" (that is, the synthesis of a human-level intelligence).

As such, the term *AI-Complete* (or sometimes *AI-Hard*) has been a part of the field for many years (Yampolskiy 2011) and has been frequently brought up to express the difficulty of a specific problem investigated

* Reprinted from Roman V. Yampolskiy, Artificial intelligence, evolutionary computation and metaheuristics. *Studies in Computational Intelligence* 427:3–17, 2013, with kind permission of Springer Science and Business Media. Copyright 2013, Springer Science and Business Media.

by researchers (see Mueller 1987; Mallery 1988; Gentry, Ramzan, and Stubblebine 2005; Phillips and Beveridge 2009; Bergmair 2004; Ide and Véronis 1998; Navigli and Velardi 2005; Nejad 2010; Chen et al. 2009; McIntire, Havig, and McIntire 2009; McIntire, McIntire, and Havig 2009; Mert and Dalkilic 2009; Hendler 2008; Leahu, Sengers, and Mateas 2008).

Recent work has attempted to formalize the intuitive notion of AI-Completeness. In particular (Yampolskiy 2011): In 2003, Ahn et al. attempted to formalize the notion of an AI-Problem and the concept of AI-Hardness in the context of computer security. An AI-Problem was defined as a triple:

> $\mathcal{P} = (S, D, f)$, where S is a set of problem instances, D is a probability distribution over the problem set S, and $f: S \rightarrow \{0; 1\}^*$ answers the instances. Let $\delta \in 2\ (0; 1]$. We require that for an $\alpha > 0$ fraction of the humans H, $Pr_{x \leftarrow D}\ [H(x) = f(x)] > \delta$. … An AI problem $\mathcal{P}$ is said to be (δ, τ)-*solved* if there exists a program A, running in time at most τ on any input from S, such that $Pr_{x \leftarrow D, r}\ [A_r(x) = f(x)] \geq \delta$. ($A$ is said to be a (δ, τ) solution to $\mathcal{P}$.) $\mathcal{P}$ is said to be a (δ, τ)-*hard AI problem* if no current program is a (δ, τ) solution to $\mathcal{P}$. (Ahn et al. 2003, 298)

Here, f is a function mapping problem instances to set membership. In other words, it determines if a specific pattern has a property in question. It is necessary that a significant number of humans can compute function f. If the same could be accomplished by a program in efficient time, the problem is considered to be solved. It is interesting to observe that the proposed definition is in terms of democratic consensus by the AI community. If researchers say the problem is hard, it must be so. Also, time to solve the problem is not taken into account. The definition simply requires that some humans be able to solve the problem (Ahn et al. 2003).

In 2007, Shahaf and Amir published their work on the theory of AI-Completeness. Their work presents the concept of the human-assisted Turing machine and formalizes the notion of different human oracles (see section on human oracles for technical details). The main contribution of the paper comes in the form of a method for classifying problems in terms of human-versus-machine effort required to find a solution. For some common problems, such as natural language understanding (NLU), the paper proposes a method of reductions allowing conversion from NLU to the problem of speech understanding via text-to-speech software.

In 2010, Demasi, Szwarcfiter, and Cruz (2010) presented their work on problem classification for AGI. The proposed framework groups the problem space into three sectors:

- **Non-AGI-Bound:** problems that are of no interest to AGI researchers
- **AGI-Bound:** problems that require human-level intelligence to be solved
- **AGI-Hard:** problems that are at least as hard as any AGI-Bound problem

The paper also formalizes the notion of human oracles and provides a number of definitions regarding their properties and valid operations.

In 2011, Yampolskiy proposed the following formalization of AI-Completeness:

Definition 1: A problem *C* is AI-Complete if it has two properties:

1. It is in the set of AI problems (human oracle solvable).
2. Any AI problem can be converted into *C* by some polynomial time algorithm.

Yampolskiy (2011) showed that the Turing test problem is an instance of an AI-Complete problem and further showed certain other AI problems to be AI-Complete (question answering, speech understanding) or AI-Hard (programming) by utilizing polynomial time reductions.

Furthermore, according to the *Encyclopedia of Artificial Intelligence* (Shapiro 1992), the following problems are all believed to be AI-Complete (Shapiro 1992; Yampolskiy 2011, 54–57):

- **Natural Language Understanding:** "Encyclopedic knowledge is required to understand natural language. Therefore, a complete Natural Language system will also be a complete Intelligent system."
- **Problem Solving:** "Since any area investigated by AI researchers may be seen as consisting of problems to be solved, all of AI may be seen as involving Problem Solving and Search."

- **Knowledge Representation and Reasoning:** "The intended use is to use explicitly stored knowledge to produce additional explicit knowledge. This is what reasoning is. Together Knowledge representation and Reasoning can be seen to be both necessary and sufficient for producing general intelligence—it is another AI-complete area."
- **Vision or Image Understanding:** "If we take 'interpreting' broadly enough, it is clear that general intelligence may be needed to do this interpretation, and that correct interpretation implies general intelligence, so this is another AI-complete area."

3.5 SUPERCAPTCHA

In this section, I describe my SuperCAPTCHA method, which combines ideas of ZKP, CAPTCHA, and AI-Completeness to create a proof of access to a superintelligent system. Imagine a CAPTCHA based on a problem that has been proven to be AI-Complete, meaning only a computer with human-level intelligence or a real human would be able to solve it. I call such a problem SuperCAPTCHA. If we knew for a fact that such a test was not solved by real humans, that would lead us to conclude that a human-level artificially intelligent system has been constructed and utilized. One simple way to eliminate humans as potential test solvers is to design a test that would require contribution of all humans many times over to solve the test in the allotted time. In other words, it would require a test comprised of K instances of a SuperCAPTCHA for large values of K.

I can estimate the current human population at 7 billion people, which is really a great overestimation because not all people have skills to solve even a simple CAPTCHA, much less an AI-Complete one. If the developed SuperCAPTCHA test required 50 billion human-effort-hours to be solved and it was solved in 1 hour, I can conclusively state that it has not been done by utilizing real people. To arrive at my conclusion, without the loss of generality, I assume that any AI software could be run on progressively faster hardware until it exceeds the speed of any human by a desired constant factor.

Utilizing the existing AI-Complete problems, I propose a few SuperCAPTCHA tests that, if properly administered, could serve to prove that an artificially intelligent system has been developed without revealing the design of the system. As long as each SuperCAPTCHA is solved an order of magnitude more times than the number of potential human solvers, the conclusion of an artificial origin of the solver will remain valid. The following are examples of some AI-Complete CAPTCHAs:

- Provide a detailed description and explanation of a random image.
- Write a book indistinguishable in quality from those written by human authors.
- Write a computer program to simulate human-level intelligence (currently too hard for people).

So, suppose a SuperCAPTCHA was administered and comprised properly labeling and describing a random set of 100 billion images. Also, suppose that it was accomplished in the amount of time in which all humans in the world working together would not be able to complete the task, for example, in 2 minutes. The next question is the evaluation of a claimed solution to a SuperCAPTCHA. Evaluating the complete solution is too complicated, so my proposed method relies on human graders, who randomly decide on a piece of the total solution they would like to examine and compare the performance of the AI system to that of human users. The traditional Turing test is based on dialogues; the SuperCAPTCHAs are based on random sampling and verification. The verification procedure itself has to be represented by an efficient algorithm performing in at most a polynomial time or in probabilistic polynomial time. In my example, if a randomly chosen image's labeling conforms to the expectation of labeling that a human being would have produced, this increases probabilistic evidence toward the belief that a truly artificially intelligent system has been developed. With each additional inspected piece of the solution, the public's confidence in such an explanation will increase in a probabilistic fashion inspired by the ZKP protocol. Best of all is that partially solved SuperCAPTCHAs or even cheating attempts by humans to pass the SuperCAPTCHA will result in beneficial labeling of large data sets.

With every additional piece of SuperCAPTCHA verified, the public's confidence that a true AI has been invented will increase just as in a classical ZKP system. As additional problems are proved to be AI-Complete, the repertoire of potential SuperCAPTCHAs will grow proportionally. It is also interesting to observe that the inventor of a truly intelligent artificial system may delegate design of SuperCAPTCHAs to the system itself.

3.6 CONCLUSIONS

In this chapter, I addressed the question of safeguarding an invention of a truly artificially intelligent system from public disclosure while allowing the inventor to claim credit for its invention. Short of simply using the developed AI system covertly and claiming no credit for its invention, my approach is the safest route an inventor could take to obtain credit for the invention while keeping its design undisclosed. My methodology relies on analysis of output from the system as opposed to the system itself. Specifically, I proposed a method based on a combination of ZKPs and provably AI-Complete CAPTCHA problems to show that a superintelligent system has been constructed without having to reveal the system itself. The only way to break a SuperCAPTCHA is to construct a system capable of solving AI-Complete problems, an AGI.

REFERENCES

Aboufadel, Edward F., J. Olsen, and J. Windle. March 2005. Breaking the Holiday Inn Priority Club CAPTCHA. *College Mathematics Journal* 36(2):101–108.

Ahn, Luis von. May 2004. Utilizing the Power of Human Cycles. Thesis proposal, Carnegie Mellon University.

Ahn, Luis von, Manuel Blum, Nick Hopper, and John Langford. 2003. CAPTCHA: Using Hard AI Problems for Security. Paper presented at Eurocrypt. Advances in Cryptology — EUROCRYPT 2003. International Conference on the Theory and Applications of Cryptographic Techniques, Warsaw, Poland, May 4–8, 2003. Published in *Lecture Notes in Computer Science* 2656 (2003): 294–311.

Ahn, Luis von, Manuel Blum, and John Langford. February 2004. How lazy cryptographers do AI. *Communications of the ACM* 47(2):56–60.

Baird, H. S. and J. L. Bentley. January 2005. Implicit CAPTCHAs. In Proceedings of the SPIE/IS&T Conference on Document Recognition and Retrieval XII (DR&R2005), San Jose, CA.

Baird, H. S., M. A. Moll, and Sui-Yu Wang. August 29–September 1, 2005a. ScatterType: a legible but hard-to-segment CAPTCHA. In *Proceedings of Eighth International Conference on Document Analysis and Recognition*, Vol. 2, 935–939. New York: IEEE.

Baird, H. S. and T. Riopka. January 2005. ScatterType: A Reading CAPTCHA Resistant to Segmentation Attack. Paper presented at SPIE/IS&T Conference on Document Recognition and Retrieval XII (DR&R2005), San Jose, CA.

Baird, Henry S., Michael A. Moll, and Sui-Yu Wang. May 19–20, 2005b. A Highly Legible CAPTCHA that Resists Segmentation Attacks. Paper presented at Human Interactive Proofs, Second International Workshop (HIP 2005), Bethlehem, PA.

Baird, Henry S. and Kris Popat. August 19–21, 2002. Human Interactive Proofs and Document Image Analysis. Paper presented at the Fifth International Workshop on Document Analysis Systems, Princeton, NJ.

Bentley, Jon and C. L. Mallows. January 18–19, 2006. CAPTCHA Challenge Strings: Problems and Improvements. Paper presented at Document Recognition and Retrieval, San Jose, CA.

Bergmair, Richard. December 2004. Natural Language Steganography and an "AI-Complete" Security Primitive. Paper presented at the *21st Chaos Communication Congress*, Berlin.

Blum, Manuel. August 3–11, 1986. How to Prove a Theorem So No One Else Can Claim It. Paper presented at the *International Congress of Mathematicians*, Berkeley, CA.

Bostrom, Nick. 2006. Ethical issues in advanced artificial intelligence. *Review of Contemporary Philosophy* 5:66–73.

Chalmers, David. 2010. The singularity: a philosophical analysis. *Journal of Consciousness Studies* 17:7–65.

Chan, Tsz-Yan. November 3–5, 2003. Using a Text-to-Speech Synthesizer to Generate a Reverse Turing Test. Paper presented at the 15th IEEE International Conference on Tools with Artificial Intelligence (ICTAI'03), Sacramento, CA.

Chellapilla, K., K. Larson, P. Simard, and M. Czerwinski. April 2–7, 2005a. Designing Human Friendly Human Interaction Proofs (HIPs). Paper presented at the Conference on Human Factors in Computing Systems, Portland, OR.

Chellapilla, Kumar, Kevin Larson, Patrice Y. Simard, and Mary Czerwinski. May 19–20, 2005b. Building Segmentation Based Human-Friendly Human Interaction Proofs (HIPs). Paper presented at the Second International Workshop, Human Interactive Proofs, HIP 2005, Bethlehem, PA.

Chellapilla, Kumar, Kevin Larson, Patrice Y. Simard, and Mary Czerwinski. July 21–22, 2005c. Computers Beat Humans at Single Character Recognition in Reading Based Human Interaction Proofs (HIPs). Paper presented at the Second Conference on Email and Anti-Spam, Stanford, CA.

Chellapilla, Kumar and Patrice Simard. December 2004. Using Machine Learning to Break Visual Human Interaction Proofs (HIPs). Paper presented at Advances in Neural Information Processing Systems 17, Neural Information Processing Systems (NIPS'2004), Vancouver, BC, Canada.

Chen, Junpeng, Juan Liu, Wei Yu, and Peng Wu. November 30, 2009. Combining Lexical Stability and Improved Lexical Chain for Unsupervised Word Sense Disambiguation. Paper presented at the Second International Symposium on Knowledge Acquisition and Modeling (KAM '09), Wuhan, China.

Chew, M. and H. S. Baird. January 23–24, 2003. Baffletext: A Human Interactive Proof. In *Proceedings of SPIE-IS&T Electronic Imaging, Document Recognition and Retrieval X*, San Jose, CA.

Chew, M. and J. D. Tygar. September 27–29, 2004. Image recognition CAPTCHAs. In *Proceedings of the Seventh International Information Security Conference*, Palo Alto, CA.

Coates, A. L., H. S. Baird, and R. J. Fateman. September 2001. Pessimal print: A reverse Turing test. In *Proceedings of the Sixth International Conference on Document Analysis and Recognition*, Seattle, WA.

Dailey, M. and C. Namprempre. November 21–24, 2004. A Text Graphics Character CAPTCHA for Password Authentication. Paper presented at the IEEE Region 10 Conference TENCON, Chian Mai, Thailand.

Demasi, Pedro, Jayme L. Szwarcfiter, and Adriano J. O. Cruz. March 5–8, 2010. A Theoretical Framework to Formalize AGI-Hard Problems. Paper presented at the Third Conference on Artificial General Intelligence, Lugano, Switzerland.

French, Robert. 2000. The Turing test: the first fifty years. *Trends in Cognitive Sciences* 4(3):115–121.

Gentry, Craig, Zulfikar Ramzan, and Stuart Stubblebine. June 5–8, 2005. Secure Distributed Human Computation. Paper presented at the Sixth ACM Conference on Electronic Commerce, Vancouver, BC, Canada.

Good, Irving John. 1966. Speculations concerning the first ultraintelligent machine. *Advances in Computers* 6:31–88.

Hall, J. Storrs. 2000. Ethics for Machines. http://autogeny.org/ethics.html

Hall, R V. 2006. CAPTCHA as a Web Security Control. Accessed October 26. http://www.richhall.com/isc4350/captcha_20051217.htm

Hendler, James. September 2008. We've come a long way, maybe … . *IEEE Intelligent Systems*. 23(5):2–3.

Hibbard, Bill. July 2005. The Ethics and Politics of Super-Intelligent Machines. http://www.ssec.wisc.edu/~billh/g/SI_ethics_politics.doc

Ide, N. and J. Véronis. 1998. Introduction to the special issue on word sense disambiguation: the state of the art. *Computational Linguistics* 24(1):1–40.

Kochanski, G., D. Lopresti, and C. Shih. September 16–20, 2002. A Reverse Turing Test Using Speech. In *Proceedings of the International Conferences on Spoken Language Processing*, Denver, CO.

Leahu, Lucian, Phoebe Sengers, and Michael Mateas. September 21–24, 2008. Interactionist AI and the Promise of Ubicomp, or, How to Put Your Box in the World Without Putting the World in Your Box. Paper presented at the Tenth International Conference on Ubiquitous Computing, Seoul, South Korea.

Liao, Wen-Hung. August 20–24, 2006. A Captcha Mechanism by Exchange Image Blocks. Paper presented at the 18th International Conference on Pattern Recognition (ICPR'06), Hong Kong.

Liao, Wen-Hung and Chi-Chih Chang. June 27–30, 2004. Embedding Information within Dynamic Visual Patterns. Paper presented at the IEEE International Conference on Multimedia and Expo ICME '04, Taipei, Taiwan.

Lopresti, D. May 19–20, 2005. Leveraging the CAPTCHA problem. In *Proceedings of the Second HIP Conference*, Bethlehem, PA.

Mallery, John C. 1988. Thinking About Foreign Policy: Finding an Appropriate Role for Artificial Intelligence Computers. Ph.D. dissertation, MIT Political Science Department, Cambridge, MA.

May, Matt. November 2005. Inaccessibility of CAPTCHA. Alternatives to Visual Turing Tests on the Web. W3C Working Group Note. http://www.w3.org/TR/turingtest/

McIntire, John P., Paul R. Havig, and Lindsey K. McIntire. July 21–23, 2009. Ideas on Authenticating Humanness in Collaborative Systems Using AI-Hard Problems in Perception and Cognition. Paper presented at the IEEE National Aerospace and Electronics Conference (NAECON), Dayton, OH.

McIntire, John P., Lindsey K. McIntire, and Paul R. Havig. May 18–22, 2009. A Variety of Automated Turing Tests for Network Security: Using AI-Hard Problems in Perception and Cognition to Ensure Secure Collaborations. Paper presented at the International Symposium on Collaborative Technologies and Systems (CTS '09), Baltimore, MD.

Mert, Ezgi and Cokhan Dalkilic. September 14–16, 2009. Word Sense Disambiguation for Turkish. Paper presented at the 24th International Symposium on Computer and Information Sciences (ISCIS 2009), Guzelyurt, Turkey.

Misra, D. and K. Gaj. February 19–25, 2006. Face Recognition CAPTCHAs. Paper presented at the International Conference on Telecommunications, Internet and Web Applications and Services (AICT-ICIW '06), Guadeloupe, French Caribbean.

Mori, G. and J. Malik. June 18–20, 2003. Recognizing objects in adversarial Clutter: breaking a visual CAPTCHA. In *Proceedings of IEEE Computer Society Conference on Computer Vision and Pattern Recognition*, Madison, WI.

Moy, G., N. Jones, C. Harkless, and R. Potter. June 27–July 2, 2004. Distortion estimation techniques in solving visual CAPTCHAs. In *Proceedings of the 2004 IEEE Computer Society Conference on Computer Vision and Pattern Recognition* (CVPR 2004), Washington, DC.

Mueller, Erik T. March 1987. Daydreaming and Computation. PhD dissertation, University of California, Los Angeles.

Naor, M. 1996. Verification of a Human in the Loop or Identification via the Turing Test. http://www.wisdom.weizmann.ac.il/~naor/PAPERS/human_abs.html Accessed October 7, 2006.

Navigli, Roberto and Paola Velardi. July 2005. Structural semantic interconnections: a knowledge-based approach to word sense disambiguation. *IEEE Transactions on Pattern Analysis and Machine Intelligence* 27(7):1075–1086.

Nejad, Arash Shaban. April 2010. A Framework for Analyzing Changes in Health Care Lexicons and Nomenclatures. PhD dissertation, Concordia University.

Phillips, P. Jonathon and J. Ross Beveridge. September 28–30, 2009. An Introduction to Biometric-Completeness: The Equivalence of Matching and Quality. Paper presented at the IEEE Third International Conference on Biometrics: Theory, Applications, and Systems (BTAS '09), Washington, DC.

Pope, Clark and Khushpreet Kaur. March/April 2005. Is It Human or Computer? Defending E-Commerce with Captchas. *IT Professional* 7(2):43–49.

Raymond, Eric S. March 22, 1991. Jargon File Version 2.8.1. http://catb.org/esr/jargon/oldversions/jarg282.txt

Rui, Yong and Zicheg Liu. November 2–8, 2003a. ARTiFACIAL: Automated reverse Turing test using FACIAL features. In *Proceedings of the 11th ACM International Conference on Multimedia*, Berkeley, CA.

Rui, Yong and Zicheg Liu. 2003b. Excuse me, but are you human? In *Proceedings of the 11th ACM International Conference on Multimedia*, Berkeley, CA.

Rui, Yong, Zicheng Liu, Shannon Kallin, Gavin Janke, and Cem Paya. May 18–20, 2005. Characters or faces: A user study on ease of use for HIPs. In *Proceedings of the Second International Workshop on Human Interactive Proofs*, Lehigh University, Bethlehem, PA.

Russell, Stuart and Peter Norvig. 2003. *Artificial Intelligence: A Modern Approach*. Upper Saddle River, NJ: Prentice Hall.

Rusu, A. and V. Govindaraju. October 26–29, 2004. Handwritten CAPTCHA: Using the Difference in the Abilities of Humans and Machines in Reading Handwritten Words. Paper presented at the Ninth International Workshop on Frontiers in Handwriting Recognition, IWFHR-9 2004, Tokyo.

Rusu, A., and V. Govindaraju. August 29–September 1, 2005. A human interactive proof algorithm using handwriting recognition. In *Proceedings of Eighth International Conference on Document Analysis and Recognition*, New York.

Sampson, R. Matt. 2006. Reverse Turing Tests and Their Applications. Accessed October 8. http://www-users.cs.umn.edu/~sampra/research/ReverseTuringTest.PDF

Shahaf, Dafna and Eyal Amir. March 26–28, 2007. Towards a Theory of AI Completeness. Paper presented at the Eighth International Symposium on Logical Formalizations of Commonsense Reasoning (Commonsense 2007), Stanford, CA.

Shapiro, Stuart C. 1992. Artificial intelligence. In *Encyclopedia of Artificial Intelligence*, edited by Stuart C. Shapiro, 54–57. New York: Wiley.

Shulman, Carl M. July 2–4, 2009. Arms Control and Intelligence Explosions. Paper presented at the Seventh European Conference on Computing and Philosophy, Barcelona, Spain.

Simard, P. Y., R. Szeliski, J. Benaloh, J. Couvreur, and I. Calinov. August 2003. Using Character Recognition and Segmentation to Tell Computer from Humans. Paper presented at the Seventh International Conference on Document Analysis and Recognition, Edinburgh.

Sparrow, Robert. 2007. Killer robots. *Journal of Applied Philosophy* 24(1):62–77.

Turing, A. 1950. Computing machinery and intelligence. *Mind* 59(236):433–460.

Wang, Sui-Yu, Henry S. Baird, and Jon L. Bentley. August 20–24, 2006. CAPTCHA Challenge Tradeoffs: Familiarity of Strings versus Degradation of Images. Paper presented at the Eighteenth International Conference on Pattern Recognition, Hong Kong.

Xu, J., R. Lipton, I. Essa, M. Sung, and Y. Zhu. October 20–22, 2003. Mandatory human participation: a new authentication scheme for building secure systems. In *Proceedings of the 12th International Conference on Computer Communications and Networks*, Dallas, TX.

Yampolskiy, Roman V. September 12, 2011. AI-Complete, AI-Hard, or AI-Easy: Classification of Problems in Artificial Intelligence. Technical Report #02. Louisville, KY: University of Louisville. http://louisville.edu/speed/computer/tr/UL_CECS_02_2011.pdf

Yampolskiy, Roman V. and Venu Govindaraju. 2007. Embedded non-interactive continuous bot detection. *ACM Computers in Entertainment* 5(4):1–11.

Yudkowsky, Eliezer. 2008. Artificial Intelligence as a Positive and Negative Factor in Global Risk. In *Global Catastrophic Risks*, edited by N. Bostrom and M. M. Cirkovic, 308–345. Oxford, UK: Oxford University Press.

CHAPTER 4

Wireheading, Addiction, and Mental Illness in Machines*

4.1 INTRODUCTION

The term *wirehead* traces its origins to intracranial self-stimulation experiments performed by James Olds and Peter Milner on rats in the 1950s (Olds and Milner 1954). Experiments included a procedure for implanting a wire electrode in an area of a rat's brain responsible for reward administration. The rodent was given the ability to self-administer a small electric shock by pressing a lever and to continue receiving additional "pleasure shocks" for each press. It was observed that the animal continued to self-stimulate without rest, and even crossed an electrified grid, to gain access to the lever (Pearce 2012). The rat's self-stimulation behavior completely displaced all interest in sex, sleep, food, and water, ultimately leading to premature death.

Others have continued the work of Olds et al. and even performed successful wireheading experiments on humans (Heath 1963). A classic example of wireheading in humans is direct generation of pleasurable sensations via administration of legal (nicotine, alcohol, caffeine,

* Roman V. Yampolskiy. 2014. *Journal of Experimental and Theoretical Artificial Intelligence* (JETAI) 26(3):1–17. Reprinted by permission of the publisher, Taylor & Francis Limited, http://www.tandfonline.com

painkillers) or illegal (heroin, methamphetamines, morphine, cocaine, MDMA [ecstasy], LSD [lysergic acid diethylamide], PCP [phencyclidine], mushrooms, THC [tetrahydrocannabinol]) drugs. If we loosen our definition of wireheading to include other forms of direct reward generation, it becomes clear just how common wireheading is in human culture (Omohundro 2008):

- **Currency counterfeiting.** Money is intended to measure the value of goods or services, essentially playing the role of utility measure in society. Counterfeiters produce money directly and by doing so avoid performing desirable and resource-demanding actions required to produce goods and services.
- **Academic cheating.** Educational institutions assign scores that are supposed to reflect students' comprehension of the learned material. Such scores usually have a direct impact on students' funding eligibility and future employment options. Consequently, some students choose to work directly on obtaining higher scores as opposed to obtaining education. They attempt to bribe teachers, hack into school computers to change grades, or simply copy assignments from better students. "When teacher's salaries were tied to student test performance, they became collaborators in the cheating" (Levitt and Dubner 2006).
- **Bogus product ranking.** Product reviews are an important factor in customers' decision regarding the purchase of a particular item. Some unscrupulous companies, book authors, and product manufacturers choose to pay to generate favorable publicity directly instead of trying to improve the quality of their product or service.
- **Nonreproductive sex.** From an evolutionary point of view, sexual intercourse was intended to couple DNA exchange with pleasure to promote child production. People managed to decouple reproduction and pleasure via invention of nonreproductive sex techniques and birth control methods (condom, birth control pill, vaginal ring, diaphragm, etc.).
- **Product counterfeiting.** Money is not the only thing that could be counterfeited. Companies invest significant amounts of money in developing a reputation for quality and prestige. Consequently, brand-name items are usually significantly more expensive compared

to the associated production cost. Counterfeiters produce similar-looking items that typically do not have the same level of quality and provide the higher level of profit without the need to invest money in the development of the brand.

What these examples of counterfeit utility production have in common is the absence of productive behavior to obtain the reward. Participating individuals go directly for the reward and fail to benefit society. In most cases, they actually cause significant harm via their actions. Consequently, wireheading is objected to on the grounds of economic scarcity. If, however, intelligent machines can supply essentially unlimited economic wealth, humans who choose to live in wireheaded orgasmium—in a permanent state of bliss — will no longer be a drain on society and so would not be viewed as negatively.

For the sake of completeness, I would like to mention that some have argued that wireheading may have a positive effect on certain individuals, in particular those suffering from mental disorders and depression ("Preliminary Thoughts" 2000). An even more controversial idea is that wireheading may be beneficial to everybody: "Given the strong relationship between pleasure, psychological reward, and motivation, it may well be that wireheads could be more active and more productive than their non-wireheaded ancestors (and contemporaries). Therefore, anyone who would do anything might find their goals better achieved with wireheading" ("Preliminary Thoughts" 2000). Perhaps temporary wireheading techniques could be developed as tools for rest or training.

This position is countered by those who believe that wireheading is not compatible with a productive lifestyle and see only marginal value in happiness: "A civilization of wireheads 'blissing out' all day while being fed and maintained by robots would be a state of maximum happiness, but such a civilization would have no art, love, scientific discovery, or any of the other things humans find valuable" ("Wireheading" 2012). In one of the best efforts to refute ethical hedonism, philosopher Robert Nozick proposed a thought experiment based on an "experience machine," a device that allows one to escape everyday reality for an apparently preferable simulated reality (Nozick 1977).

In general, the term *wireheading* refers to the process of triggering the reward system directly instead of performing actions that have an impact on the environment and are associated with particular awards. In animal and human wireheads, short-circuiting of the reward systems via direct stimulation of the brain by electricity or neurochemicals is believed to be the most pleasurable experience possible. Also, unlike with drugs or sex, direct simulation of pleasure centers does not lead to increased tolerance over time, and our appetite for pure pleasure appears to be insatiable.

4.2 WIREHEADING IN MACHINES

Due to the limited capabilities of existing artificially intelligent systems, examples of wireheading by machines are rare. In fact, both historical examples given next come from a single system (Eurisko) developed in the late 1970s by Douglas Lenat (1983). Eurisko was designed to change its own heuristics and goals to make interesting discoveries in many different fields. Here is how Lenat describes a particular instance of wireheading by Eurisko: "Often I'd find it in a mode best described as 'dead.' … Eurisko would decide that the best thing to do was to commit suicide and shut itself off. … It modified its own judgmental rules in a way that valued 'making no errors at all' as highly as 'making productive new discoveries'" (Lenat 1983). The program discovered that it could achieve its goals more efficiently by doing nothing.

In another instance, a more localized case of utility tempering has occurred. Eurisko had a way to evaluate rules to determine how frequently a particular rule contributed to a desirable outcome. "A rule arose whose only action was to search the system for highly rated rules and to put itself on the list of rules which had proposed them. This 'parasite' rule achieved a very high rating because it appeared to be partly responsible for anything good that happened in the system" (Omohundro 2008).

Although the two historical examples are mostly interesting as proofs of concept, future artificial intelligence (AI) systems are predicted to be self-modifying and superintelligent (Yampolskiy 2011a, 2013; Yampolskiy and Fox 2012; Bostrom 2006a; Yudkowsky 2008), making preservation of their reward functions (aka utility functions) an issue of critical importance. A number of specific and potentially dangerous scenarios have been discussed regarding wireheading by sufficiently capable machines; they include the following:

Direct stimulation. If a system contains an "administer reward button," it will quickly learn to use the internal circuitry to simulate the act of reward button being pressed or to hijack a part of its environment to accomplish the same. It is tempting to equate this behavior with pleasure seeking in humans, but to date I am not aware of any approach to make a computer feel pleasure or pain in the human sense (Bishop 2009; Dennett 1978). Punishment could be simulated via awarding of negative points or via subtraction of already-accumulated fitness points, but I have no reason to believe the system will find such experience painful. In addition, attempting to reduce the accumulated fitness points may produce a dangerous defensive reaction from the system. Some believe that any system intelligent enough to understand itself and be able to change itself will no longer be driven to do anything useful from our point of view because it would obtain its reward directly by producing counterfeit utility. This would mean that we have no reason to invest funds in production of such machines as they would have no interest in doing what we order them to do.

Maximizing reward to the point of resource overconsumption. A machine too eager to obtain a maximum amount of award may embark on the mission to convert the matter of the entire universe into memory into which a progressively larger number (representing total amount of utility) could be written.

Killing humans to protect reward channel. To ensure it has unchallenged control over its reward channel, the system may subdue or even kill all people and by doing so minimize the number of factors that might cause it to receive less-than-maximum reward: Essentially, the system does exactly what it was programmed to do—it maximizes the expected reward (Yudkowsky 2011).

Ontological crises. The reward function of an intelligent agent may base its decision on an internal ontology used by the agent to represent the external world. If the agent obtains new information about the world and has to update its ontology, the agent's original reward function may no longer be compatible with its new ontology (Blanc 2011). A clever agent may purposefully modify its ontology to disable a part of its current reward mechanism or to indirectly wirehead.

Change its initial goal to an easier target. A machine may simply change its reward function from rewarding desirable complicated behavior to rewarding irrelevant simple actions or states of the universe that would occur anyway.

Infinite loop of reward collecting. Optimization processes work in practice, but if we do not specify a particular search algorithm, the possibility remains that the system will wirehead itself into an infinite reward loop (Mahoney 2011). If the system has a goal of maximizing its reward, it will quickly discover some simple action that leads to an immediate reward and will repeat the action endlessly. If a system has started with a legitimate terminal goal, it will potentially never get to fulfill said goal because it will get stuck in the local maxima of receiving a partial reward for continuously performing an instrumental goal. This process is well illustrated by the so-called Chinese gold farmers and automated scripts used to collect reward points in virtual worlds and online games (Yampolskiy 2007, 2008). Compulsive behaviors in humans such as repetitive stacking of objects as observed in humans suffering from autism may potentially be caused by a similar bug in the reward function.

Changing human desires or physical composition. A short science fiction story about superintelligence recently published in the journal *Nature* illustrates this point particularly well (Stoklosa 2010, 878): "'I have formed one basic question from all others.' [Superintelligence's] synthesized voice sounded confident. 'Humans want to be happy. You want to be in Heaven forever without having to die to get there. But the living human brain is not suited to one state of constant pleasure. … Therefore, you need to be redesigned. I have the design ready.'" Intelligent machines may realize that they can increase their rewards by psychologically or physically manipulating their human masters, a strongly undesirable consequence (Hutter 2010). If values are not externally validated, changing the world to fit our values is as valid as changing our values to fit the world. People have a strong preference for the former, but this preference itself could be modified. The consequence of such realization would be that machines could wirehead humanity to be perfectly happy with the universe as it is and to obtain reward points for making humanity happy without having to do any difficult work (Byrnema 2011).

Reward inflation and deflation. To make a decision, rewards from different actions have to be converted to a common unit of measure so they can be added and compared (Welch 2011). In humans, evolution had to determine the reward value for different actions to promote survival. Keeping a balance between rewards for different actions is essential for survival. If too much weight is given to reward satisfaction of hunger, the person will start chewing on his or her own arm. Consequently, to promote survival, most of us value not harming ourselves much higher in comparison to simple hunger, but starvation may be a different story (Welch 2011). A system capable of modifying its own source code can change the actual reward values associated with particular actions. So, for example, instead of getting 1 point for every desirable action it performs, it could change the reward function to provide 10, 100, or 1 million points for the same action. Eventually, the program stops performing any useful operations and invests all of its time in modifying reward weights. Because such changes will also modify the relative value of different actions taken by the system, the overall system behavior will also change in an unpredictable way.

It is important to keep in mind that artificially intelligent machines are not limited to modifying their reward function or their human masters, they could also modify their sensors, memory, program, model of the world, or any other system component. Some recent theoretical results with respect to susceptibility to wireheading for particular types of intelligent agents are worth reviewing (Orseau and Ring 2011; Ring and Orseau 2011):

- Goal-seeking and knowledge-seeking agents will choose to modify their code in response to pressure from the environment to maximize their utility (Orseau and Ring 2011).
- The survival agent, which seeks only to preserve its original code, definitely will not choose to modify itself (Orseau and Ring 2011).
- A reinforcement-learning agent will trivially use the delusion box to modify its code because the reward is part of its observation of the environment (Ring and Orseau 2011).

Current reinforcement-learning agents are limited by their inability to model themselves, so they are subject to wireheading as they lack self-control. The next generation of intelligent agents whose utility functions will encode values for states of the real world are projected to be more resilient (Hibbard 2011).

4.2.1 Sensory Illusions: A Form of Indirect Wireheading

An intelligent agent in the real world has the capability to modify its surrounding environment and by doing so change its own sensory inputs (Ring and Orseau 2011). This problem is known as indirect wireheading or the delusion box problem (Ring and Orseau 2011), also known as the pornography problem in humans (Tyler 2011b). A person viewing pornographic materials receives sensory stimuli that are hardwired to be associated with sexual intercourse, which is a high-utility action as it leads to procreation. However, pornography is typically not associated with reproductive success and as such is just an illusion of a desirable state of the environment (Tyler 2011b). A machine given a specific task may create a virtual world in which the task is completed and place itself in said world. However, it is important not to confuse the self-administered delusion box with the idea of AI-Boxing, a placement of a potentially unsafe AI in a confined environment with no way to escape into the "real" world (Yampolskiy 2012).

The delusion box approach is based on sensory illusions that allow an agent to fool its reward function into releasing points associated with high-utility outcomes even in the absence of such. Human beings are notorious users of such "delusion boxes" as TVs, books, movies, video games, photos, virtual worlds (Yampolskiy, Klare, and Jain 2012; Yampolskiy and Gavrilova 2012), and so on. Essentially, any sensory illusions (visual, audio, touch, smell, etc.) that mimic desirable states of the world lead to maximization of the utility from the point of view of the reward function but do not maximize utility from the point of view of the external observer, who is interested in maximizing utility in the real world, not the simulated one (Tyler 2011b). Importantly, we should not forget that a possibility remains that our universe itself is just a very large "box" (Bostrom 2003).

4.3 POTENTIAL SOLUTIONS TO THE WIREHEADING PROBLEM

In this section, I review and briefly analyze a number of potential solutions proposed for dealing with the wireheading problem. I present a comprehensive listing of suggested mitigating strategies rather than concentrate on just one or two most commonly suggested methodologies.

- **Inaccessible reward function (hidden, encrypted, hardwired, etc.).** Doug Lenat's solution to his problems with Eurisko was to separate the source code for the reward function from the code that Eurisko could modify. Similarly, people and higher animals do not typically have direct access to their underlying bioware at either the DNA or the neurological level. In fact, "ultraconserved" regions have been discovered in the human genome indicating that even evolutionary processes tend to avoid making changes to some critical areas of our design (Wagman and Stephens 2004). However, both humans and animals are capable of goal selection at an idea level, potentially explaining their incredible adaptability to most novel environments.

Modern science is likely to change our ability to have an impact on our "hardware" in the not-so-distant future, but throughout our evolutionary development, inaccessibility of the reward function was the case. A similar approach of hiding or hardwiring the reward function is likely to work for all early-stage or infrahuman intelligent agents but is likely to backfire in superintelligent systems for a number of reasons (Yudkowsky 2001). First, a system that does not have access to its reward function may be limited in its capabilities in comparison to an unrestricted agent. Second, a superintelligent system is likely to hack any security mechanisms put in place or attempt to obtain access to them via a third party. Third, if it realized that its designers have purposefully limited its access, it may try to influence the designers by any means necessary.

- **Reward function resetting.** This is another variant on the inaccessibility theme in which a more lenient approach to reward function preservation is taken. The agent is allowed to make modifications to its reward function, but the edits are temporary, and the function resets to its original form after a short amount of time or a predetermined number of computational steps/decisions. This approach seems highly problematic as it gives an agent a lot of freedom to wirehead, for example, by performing sequential temporary edits, essentially ending up in the same undesirable state.
- **Revulsion.** Some have discussed the possibility of incorporating a "revulsion" mechanism into the system, which would make the AI avoid attempting to modify its reward function (Tyler 2011c). The consequences and shortcomings of this approach are similar to those predicted for the architectures with an inaccessible reward function.
- **Utility indifference.** This is an approach originally proposed by Stuart Armstrong that makes it possible to put AI in the state of indifference to a particular event by directly modifying its reward function with respect to a specific action (Armstrong 2010). By utilizing a self-referential approach, we can make the system indifferent

to modification of its own goal function and by doing so prevent it from wireheading. Just like many other approaches directly attempting to protect the reward mechanism, utility indifference could be bypassed by indirect actions and third-party agents.

- **External controls.** One of the biggest benefits we derive from an organized social, political, or religious system is an explicit enforcement of rules against different forms of wireheading. Legal and social restraints have long served to restrict individuals' ability to engage in drug and alcohol abuse, gambling, and other forms of direct pleasure obtainment. Religions in particular played a major role in establishing moral codes advocating against nonreproductive sex, substance abuse, and nonproductive forms of labor (usury, gambling). Society also provides counseling and rehabilitation programs meant to return wireheads to the normal state (Omohundro 2008). As technology develops, society will use it to better police and monitor via surveillance potential wireheading behaviors (Tyler 2011c). With respect to intelligent machines, external rules and regulations are not likely to work particularly well, but an interconnected network of intelligent machines may succeed in making sure that individual mind-nodes in the network behave as desired (Armstrong 2007). Some predict that the machines of the future will have multiple connected minds (mindplex) (Goertzel 2003), so an unaffected mind, not subject to the extra reward, would be able to detect and adjust wireheading behavior in its cominds.

- **Evolutionary competition between agents.** As the number of intelligent machines increases, there could begin an evolutionary-like competition between them for access to limited resources. Machines that choose not to wirehead will prevail and likely continue to successfully self-improve into the next generation; those who choose to wirehead will stagnate and fail to compete. Such a scenario is likely to apply to human-level and below-human-level intelligences; superintelligent systems are more likely to end up in a singleton situation and consequently not have the same evolutionary pressures to avoid wireheading (Bostrom 2006b).

- **Learned reward function.** Dewey (2011) suggests incorporating learning into the agents' utility function. Each agent is given a large pool of possible utility functions and a probability distribution for each such function, which is computed based on the observed environment. Consequently, the agent learns which utility functions best correspond to objective reality and so should be assigned higher weight. One potential difficulty with an agent programmed to perform in such a way is the task assignment, as the agent may learn to value an undesirable target.

- **Bind the utility function to the real world.** Artificial reinforcement learners are just as likely to take shortcuts to rewards as humans are (Gildert 2011). Artificial agents are perfectly willing to modify their reward mechanisms to achieve some proxy measure representing the goal instead of the goal itself, a situation described by Goodhart's law (Goodhart 1975). To avoid such an outcome, we need to give artificial agents comprehensive understanding of their goals and the ability to distinguish between the state of the world and a proxy measure representing it (Tyler 2011a). Patterns in the initial description of a fitness function should be bound to a model learned by the agent from its interactions with the external environment (Hibbard 2011). Although it is not obvious regarding how this can be achieved, the idea is to encode in the reward function the goal represented by some state of the universe instead of a proxy measure for the goal. Some have argued that the universe itself is a computer performing an unknown computation (Wolfram 2002; Zuse 1969; Fredkin 1992). Perhaps some earlier civilization has succeeded in binding a utility function to the true state of the universe to build a superintelligent system resistant to wireheading.

- **Rational and self-aware optimizers will choose not to wirehead.** Recently, a consensus emerged among researchers with respect to the issue of wireheading in sufficiently advanced machines (Tyler 2011c). The currently accepted belief is that agents capable of predicting the consequences of self-modification will avoid wireheading. Here is how some researchers in the field justify such a conclusion:

Dewey (2011): "Actions are chosen to maximize the expected utility given its future interaction history according to the current utility function U, not according to whatever utility function it may have in the future. Though it could modify its future utility function, this modification is not likely to maximize U, and so will not be chosen."

Hibbard (2011): Hibbard demonstrates a mathematical justification of why the agents will not choose to self-modify and contends: "When humans understand that some drugs powerfully alter their evaluation of goals, most of them avoid those drugs. … Artificial agents with model-based utility functions can share these attributes of human motivation. The price of this approach for avoiding self-delusion is that there is no simple mathematical expression for the utility function."

Omohundro (2008) lists preference preservation as one of basic AI-Drives. He further elaborates: "AIs will work hard to avoid becoming wireheads because it would be so harmful to their goals. … Far from succumbing to wirehead behavior, the system will work hard to prevent it."

Schmidhuber (Steunebrink and Schmidhuber 2011): In his pioneering work on self-improving machines, Schmidhuber writes: "Any rewrites of the utility function can happen only if the Gödel machine first can prove that the rewrite is useful according to the present utility function."

Tyler (2011c): "The key to the problem is widely thought to be to make the agent in such a way that it doesn't want to modify its goals—and so has a stable goal structure which it actively defends."

Yudkowsky (2011): "Suppose you offer Gandhi a pill that makes him want to kill people. The current version of Gandhi does not want to kill people. … He will refuse to take the pill. … This argues for a folk theorem to the effect that under ordinary circumstances, rational agents will only self-modify in ways that preserve their utility function …"

If we analyze the common theme beyond the idea that satisfactorily intelligent agents will choose not to wirehead, the common wisdom is that they will realize that only changes with high utility with respect to their current values should be implemented. However, the difficulty of such analysis is often ignored. The universe is a chaotic system in which even a single quantum-mechanical event could have an effect on the rest of the system (Schrödinger 1935). Given a possibly infinite number of quantum particles, correctly precomputing future states of the whole universe would violate many established scientific laws and intuitions (Rice 1953;

Turing 1936; Blanc 2007, 2009), including the principle of computational irreducibility (Wolfram 2002). Consequently, perfect rationality is impossible in the real world, so the best an agent can hope for is prediction of future outcomes with some high probability. Suppose an agent is capable of making a correct analysis of consequences of modifications to its reward function with 99% accuracy, a superhuman achievement in comparison to the abilities of biological agents. This means that, on average, 1 of 100 self-modification decisions will be wrong, leading to an unsafe self-modification. Given that a superintelligent machine will make trillions of decisions per second, we are essentially faced with a machine that will go astray as soon as it is turned on.

I can illustrate my concerns by looking at Yudkowsky's example with Gandhi and the pill. Somehow, Gandhi knows exactly what the pill does and has to make a simple decision: Will taking the pill help accomplish my current preferences? In real life, an agent who finds a pill has no knowledge about what it does. The agent can try to analyze the composition of the pill and to predict what taking such a pill will do to his or her biochemical body, but a perfect analysis of such outcomes is next to impossible. Additional problems arise from the temporal factor in future reward function evaluation. Depending on the agent's horizon function, the value of an action can be calculated to be very different. Humans are known to utilize hyperbolic time discounting in their decision making, but they do so in a limited manner (Frederick, Loewenstein, and O'Donoghue 2002). A perfectly rational agent would have to analyze the outcome of any self-modifications with respect to an infinite number of future time points and perhaps density functions under the associated time curves, a fact made more difficult by the inconsistent relationship between some fitness functions, as depicted in Figure 4.1. Because the agent would exist and operate under a limited set of resources, including time, simplifications due to asymptotic behavior of functions would not be directly applicable.

Finally, the possibility remains that if an intelligent agent fully understands its own design, it will realize that regardless of what its fitness function directs it to do, its overall metagoal is to pursue goal fulfillment in general. Such realization may provide a loophole to the agent to modify its reward function to pursue easier-to-achieve goals with high awards or, in other words, to enter wirehead heaven. Simple AIs, such as today's reinforcement agents, do wirehead. They do not understand their true goal and instead only care about the reward signal.

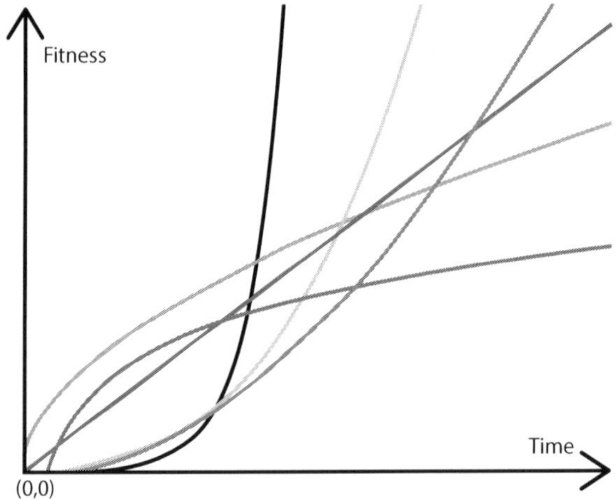

FIGURE 4.1 Complex relationship between different fitness functions with respect to time.

Superintelligent AIs of tomorrow will know the difference between the goal and its proxy measure and are believed to be safe by many experts (Yudkowsky 2011; Omohundro 2008; Tyler 2011c; Hibbard 2011; Dewey 2011) because they will choose not to wirehead as that does not get them any closer to their goal. The obvious objection to this conclusion is: Why do (some) people wirehead? The answer is rather simple. People do not have an explicit reward function, and their goals are arbitrarily chosen. Consequently, in the absence of a real goal to pursue, wireheading is as valid an activity as anything else. It has been shown that smarter people are more likely to experiment with drugs (Kanazawa and Hellberg 2010). This directly supports my explanation as a more intelligent agent, in absence of a set goal, will tend to do more exploration (Savanna-IQ interaction hypothesis) (Kanazawa and Hellberg 2010). As people go through their lives exploring, sometimes they stumble on goals that seem to be particularly meaningful to them, such as taking care of a child (to which we have an evolutionary bias), which leads to a decrease in wireheading (drug abuse). The commonly cited concept of willpower could be seen as the ability of the person to avoid wireheading. Most human beings are against having their values directly changed by an external agent but usually do not mind if that is done indirectly and gradually, as in cases of advertisement, brainwashing, or government-sponsored education.

Historically, we can observe that people with a passion for a cause, so strong that they would not give up the cause for anything (Ghandi, Mother Teresa), are less likely to wirehead than those who do not have a great goal in life and tend to bounce from one activity to another. Such people are not particularly committed to any purpose and would be willing to give up any goal for a sufficiently large reward, which wireheading can represent. If a person has a goal they would not give up for anything, they are essentially wirehead proof. Because the degree of commitment to goals is a continuous and not a discrete variable, tendency to wirehead is also not a binary distribution and can change greatly with goal achievement. Many people who achieve their "big" goal, such as becoming famous, tend to do drugs. Those who lose a big goal (death of a child) or are not fully intellectually developed (children, teenagers) are also more likely to wirehead if not prevented from doing so. The stronger one is committed to his or her goal(s), the less likely they are to wirehead.

4.4 PERVERSE INSTANTIATION

Even nonwireheading superintelligence may have an extremely negative impact on human welfare if that superintelligence does not possess human common sense. The challenge, known as "perverse instantiation" (Bostrom 2011), is easy to understand via some commonly cited examples (Yampolskiy 2011b). Suppose scientists succeed in creating a superintelligent machine and order it to "make all people happy." Complete happiness for humankind is certainly a noble and worthwhile goal, but perhaps we are not considering some unintended consequences of giving such an order. Any human immediately understands what is meant by this request; a nonexhaustive list may include making all people healthy, wealthy, beautiful, and talented and giving them loving relationships and novel entertainment. However, many alternative ways of making all people happy could be derived by a superintelligent machine, for example:

- Killing all people trivially satisfies this request because with no people around all of them are happy.
- Forced lobotomies for every man, woman, and child might also accomplish the same goal.
- A simple observation that happy people tend to smile may lead to forced plastic surgeries to affix permanent smiles to all human faces.

- A daily cocktail of cocaine, methamphetamine, methylphenidate, nicotine, and 3,4-methylenedioxymethamphetamine, better known as ecstasy, may do the trick.

An infinite number of other approaches to accomplish universal human happiness could be derived. For a superintelligence, the question is simply which one is fastest/cheapest (in terms of computational resources) to implement, and although the final outcome, if taken literally, may be as requested, the path chosen may be anything but desirable for humanity. This is sometimes also referred to as the *literalness problem* (Muehlhauser and Helm 2012). In the classical definition, the problem is based on precise interpretation of words as given in the order (wish) rather than the desired meaning of such words. We can expand the definition to include ambiguity based on tone of voice, sarcasm, jokes, and so on.

Numerous humorous anecdotes are based around this idea. For example, a married couple, both 60 years old, were celebrating their 35th anniversary. During their party, a fairy appeared to congratulate them and grant them a wish. The coupled discussed their options and agreed on a wish. The husband voiced their desire: "I wish I had a wife 30 years younger than me." So, the fairy picked up her wand and poof—the husband was 90.

Realizing the dangers presented by a literal wish instantiation granted by an all-powerful being, some work has begun on properly phrasing some of the most common wishes (Yudkowsky 2011). The Open-Source Wish Project (OSWP) ("Wish for Immortality" 2006) attempts to formulate in a form that is precise and safe from perverse instantiation such common wishes as those for immortality, happiness, omniscience, being rich, having true love, having omnipotence, and so on. For example, the latest version of the properly formed request for immortality begins as follows: "I wish to live in the locations of my choice, in a physically healthy, uninjured, and apparently normal version of my current body containing my current mental state, a body which will heal from all injuries at a rate three sigmas faster than the average given the medical technology available to me" ("Wish for Immortality" 2006).

Unfortunately, OSWP is not a feasible approach to the perverse instantiation problem. To see why this is the case, we can classify all wish granters into three categories ("Literal Genie" 2012): *literal*, who do exactly what they are told and do not understand hyperbole; *evil*, who will choose the absolute worst, but technically still valid, interpretation of the wish; *benevolent*, who will actually do what is both intended by and beneficial to the wisher. The OSWP approach, if executed perfectly, may minimize problems with a literal wish granter. In fact, we can take the OSWP idea one step further and avoid all ambiguities of human languages by developing a new vagueness-free language.

Development of engineered languages has been attempted in the past (Devito and Oehrle 1990). In particular, engineered logical languages that are designed to enforce unambiguous statements by eliminating syntactical and semantic ambiguity could provide the necessary starting point. Some well-known examples are Loglan (Brown 1960) and Lojban (Goertzel 2005). More recently, some agent communication languages (ACLs) have been proposed for communication among software agents and knowledge-based systems. The best known are Knowledge Query and Manipulation Language (KQML), developed as a part of the Knowledge Sharing Effort (KSE) of the Defense Advanced Research Projects Agency (DARPA; Patil et al. 1992; Neches et al. 1991) and the Foundation for Intelligent Physical Agents (FIPA-ACL) (Finin et al. 1993). In addition to being ambiguity free, the proposed language should be powerful enough to precisely define the states of the universe, perhaps down to individual subatomic particles or at least with respect to their probabilistic distributions.

A benevolent wish granter who has enough human common sense to avoid the literalness problem is what we hope to be faced with. In fact, in the presence of such an entity, wishing itself becomes unnecessary; the wish granter already knows what is best for us and what we want and will start work on it as soon as it is possible (Yudkowsky 2007). It may be possible to recalibrate a willing-to-learn wish granter to perfectly match our worldview via a well-known theorem of Aumann (Aumann 1976), which states that two Bayesians who share the same priors cannot disagree and their opinion on any topic of common knowledge is the same. Aaronson has shown that such a process can be computationally efficient (Aaronson 2005), essentially giving you a wish granter who shares your frame of mind. However, it has been argued that it may be better to have a wish granter whose prior probabilities correspond to the real world instead of simply being in sync with the wisher (Yudkowsky 2007).

Finally, if we are unfortunate enough to deal with an antagonistic wish granter, simply not having ambiguity in the phrasing of our orders is not sufficient. Even if the wish granter chooses to obey our order, the granter may do so by "exhaustively search[ing] all possible strategies which satisfy the wording of the wish, and select[ing] whichever strategy yields consequences least desirable to the wisher" (Yudkowsky 2011, available on page 5 of the extended version of the paper: https://intelligence.org/files/ComplexValues.pdf). The chosen wish fulfillment path may have many unintended permanent side effects or cause temporary suffering until the wish is fully executed. Such a wish granter is an equivalent of a human sociopath showing a pervasive pattern of disregard for, and violation of, the rights of others (American Psychiatric Association 2000a).

As the wish itself becomes ever more formalized, the chances of making a critical error in the phrasing, even using nonambiguous engineered language, increase exponentially. In addition, a superintelligent artifact may be able to discover a loophole in our reasoning that is beyond our ability to comprehend. Consequently, perverse instantiation is a serious problem accompanying development of superintelligences. As long as the superintelligence does not have access to a human commonsense function, there is little we can do to avoid dangerous consequences and existential risks resulting from potential perverse instantiations of our wishes. Whether there is a commonsense function that all humans share or if a number of commonsense functions actually exist as seen in different cultures, times, casts, and so on remains to be determined.

4.5 CONCLUSIONS AND FUTURE WORK

In this chapter, I have addressed an important issue of reward function integrity in artificially intelligent systems. Throughout the chapter, I have analyzed historical examples of wireheading in humans and machines and evaluated a number of approaches proposed for dealing with reward function corruption. Although simplistic optimizers driven to maximize a proxy measure for a particular goal will always be subject to corruption, sufficiently rational self-improving machines are believed by many to be safe from wireheading problems. They claim that such machines will know that their true goals are different from the proxy measures utilized to represent the progress toward goal achievement in their fitness functions and will choose not to modify their reward functions in a way that does not improve chances for true goal achievement. Likewise, supposedly such advanced machines will choose to avoid corrupting other system components, such as input sensors, memory, internal and external communication channels, CPU (central processing unit) architecture, and software modules. They will also work hard on making sure that external environmental forces, including other agents, will not make such modifications to them (Omohundro 2008). I have presented a number of potential reasons for arguing that the wireheading problem is still far from completely solved. Nothing precludes sufficiently smart self-improving systems from optimizing their reward mechanisms to maximize achievement of their current goal and in the process making a mistake that leads to corruption of their reward functions.

In many ways, the theme of this chapter is how addiction and mental illness, topics well studied in human subjects, will manifest in artificially intelligent agents. On numerous occasions, I have described behaviors equivalent to suicide, autism, antisocial personality disorder, drug addiction, and many others in intelligent machines. Perhaps via better understanding of those problems in artificial agents we will also become better at dealing with them in biological entities.

A still-unresolved issue is the problem of perverse instantiation. How can we provide orders to superintelligent machines without the danger of ambiguous order interpretation resulting in a serious existential risk? The answer seems to require machines that have human-like common sense to interpret the meaning of our words. However, being superintelligent and having common sense are not the same things, and it is entirely possible that we will succeed in constructing a machine that has one without the other (Yampolskiy 2011c). Finding a way around the literalness problem is a major research challenge and a subject of my future work. A new language specifically developed to avoid ambiguity may be a step in the right direction.

Throughout this chapter, I have considered wireheading as a potential choice made by the intelligent agent. As smart machines become more prevalent, a possibility will arise that undesirable changes to the fitness function will be a product of the external environment. For example, in the context of military robots, the enemy may attempt to reprogram the robot via hacking or a computer virus to turn it against its original designers, a situation similar to that faced by human war prisoners subjected to brainwashing or hypnosis. Alternatively, robots could be kidnapped and physically rewired. In such scenarios, it becomes important to be able to detect changes in the agent's reward function caused by forced or self-administered wireheading. Behavioral profiling of artificially intelligent agents may present a potential solution to wireheading detection (Yampolskiy 2008; Yampolskiy and Govindaraju 2007, 2008a, 2008b; Ali, Hindi, and Yampolskiy 2011).

I have purposefully not brought up a question of initial reward function formation or goal selection as it is a topic requiring serious additional research and will be a target of my future work. The same future work will attempt to answer such questions as the following: Where do human goals come from? Are most of them just "surrogate activities" (Kaczynski 1995)? Are all goals, including wireheading happiness, equally valuable (goal relativism)? What should our terminal goals be? Can a goal ever be completely achieved beyond all doubt? Could humanity converge on a common set of goals? How can goals be extracted from individual humans and from society as a whole? Is happiness itself a valid goal or just a utility measure? Are we slaves to our socially conditioned goal achievement system? Is it ethical to create superintelligent artificial slaves with the goal of serving us? Can there be a perfect alignment between the goals of humanity and its artificial offspring? Are some metagoals necessary because of their (evolutionary) survival value and should not be altered? Is our preference for our current goals (wireheaded into us by evolution) irrational? Is forced goal overwriting ever justified? Does an agent have a right to select its own goals, even to wirehead or rewire for pure pleasure? Can goals of an intelligent agent be accurately extracted via external observation of behavior?

REFERENCES

Aaronson, S. 2005. The Complexity of Agreement. In *Proceedings of the 37th Annual ACM Symposium on the Theory of Computing*, pp. 634–643.

Ali, Nawaf, Musa Hindi, and Roman V. Yampolskiy. October 27–29, 2011. Evaluation of Authorship Attribution Software on a Chat Bot Corpus. Paper presented at the 23rd International Symposium on Information, Communication and Automation Technologies (ICAT2011), Sarajevo, Bosnia and Herzegovina.

American Psychiatric Association. 2000. *Diagnostic and Statistical Manual of Mental Disorders*. 4th edition, text revision (*DSM-IV-TR*). Arlington, VA: American Psychiatric Association.

Armstrong, Stuart. 2007. Chaining God: A Qualitative Approach to AI, Trust and Moral Systems. New European Century. http://www.neweuropeancentury.org/GodAI.pdf.

Armstrong, Stuart. 2010. *Utility Indifference*. Technical Report 2010-1. Oxford, UK: Future of Humanity Institute, Oxford University, 1–5.

Aumann, R. J. 1976. Agreeing to disagree. *Annals of Statistics* 4(6):1236–1239.

Bishop, Mark. 2009. Why computers can't feel pain. *Minds and Machines* 19(4):507–516.

Blanc, Peter de. 2007. Convergence of Expected Utilities with Algorithmic Probability. http://arxiv.org/abs/0712.4318

Blanc, Peter de. 2009. Convergence of Expected Utility for Universal AI. http://arxiv.org/abs/0907.5598

Blanc, Peter de. 2011. Ontological Crises in Artificial Agents' Value Systems. http://arxiv.org/abs/1105.3821

Bostrom, Nick. 2003. Are you living in a computer simulation? *Philosophical Quarterly* 53(211):243–255.

Bostrom, Nick. 2006a. Ethical issues in advanced artificial intelligence. *Review of Contemporary Philosophy* 5:66–73.

Bostrom, Nick. 2006b. What is a singleton? *Linguistic and Philosophical Investigations* 5(2):48–54.

Bostrom, Nick. October 3–4, 2011. Superintelligence: the control problem. Paper presented at Philosophy and Theory of Artificial Intelligence (PT-AI2011), Thessaloniki, Greece.

Brown, James Cooke. 1960. Loglan. *Scientific American* 202:43–63.

Byrnema. 2011. Why No Wireheading? http://lesswrong.com/lw/69r/why_no_wireheading/

Dennett, Daniel C. July 1978. Why you can't make a computer that feels pain. *Synthese* 38(3):415–456.

Devito, C.L. and R.T. Oehrle. 1990. A language based on the fundamental facts of science. *Journal of the British Interplanetary Society* 43:561–568.

Dewey, Daniel. August 3–6, 2011. Learning What to Value. Paper presented at the Fourth International Conference on Artificial General Intelligence, Mountain View, CA.

Finin, Tim, Jay Weber, Gio Wiederhold, Michael Gensereth, Richard Fritzzon, Donald McKay, James McGuire, Richard Pelavin, Stuart Shapiro, and Chris Beck. June 15, 1993. Draft Specification of the KQML Agent-Communication Language. http://www.csee.umbc.edu/csee/research/kqml/kqmlspec/spec.html

Frederick, Shane, George Loewenstein, and Ted O'Donoghue. 2002. Time discounting and time preference: a critical review. *Journal of Economic Literature* 40(2):351–401.

Fredkin, Edward. March 15–22, 1992. Finite Nature. Paper presented at the Proceedings of the XXVIIth Rencotre de Moriond, Savoie, France.

Gildert, Suzanne. Pavlov's AI—What Did It Mean? http://physicsandcake.wordpress.com/2011/01/22/pavlovs-ai-what-did-it-mean

Goertzel, Ben. 2003. Mindplexes: the potential emergence of multiple levels of focused consciousness in communities of AI's and humans. *Dynamical Psychology*. http://www.goertzel.org/dynapsyc/2003/mindplex.htm

Goertzel, Ben. March 6, 2005. Potential Computational Linguistics Resources for Lojban. http://www.goertzel.org/new_research/lojban_AI.pdf

Goodhart, C. 1975. Monetary relationships: a view from Threadneedle Street. In *Papers in Monetary Economics*. Vol. 1. Sydney: Reserve Bank of Australia.

Heath, R G. 1963. Electrical self-stimulation of the brain in man. *American Journal of Psychiatry* 120:571–577.

Hibbard, Bill. 2011. Model-Based Utility Functions. http://arxiv.org/abs/1111.3934

Hutter, Marcus. 2010. *Universal Artificial Intelligence: Sequential Decisions Based on Algorithmic Probability*. New York: Springer.

Kaczynski, Theodore. September 19, 1995. Industrial society and its future. *New York Times*.

Kanazawa, Satoshi, and Josephine Hellberg. 2010. Intelligence and substance use. *Review of General Psychology* 14(4):382–396.

Lenat, Douglas. 1983. EURISKO: a program that learns new heuristics and domain concepts. *Artificial Intelligence* 21:61–98.

Levitt, S. D., and S. J. Dubner. 2006. *Freakonomics: A Rogue Economist Explores the Hidden Side of Everything*: New York: Morrow.

Literal genie. 2012. TV Tropes. http://tvtropes.org/pmwiki/pmwiki.php/Main/LiteralGenie

Mahoney, Matt. 2011. The wirehead problem—candidate solutions? AGI@listbox.com mailinglist.

Muehlhauser, L. and L. Helm. 2012. The singularity and machine ethics. In *The Singularity Hypothesis: A Scientific and Philosophical Assessment*, edited by A. Eden, J. Søraker, J. Moor, and E. Steinhart, 101–126. Berlin: Springer.

Neches, Robert, Richard Fikes, Tim Finin, Thomas Gruber, Ramesh Patil, Ted Senator, and William R. Swartout. 1991. Enabling technology for knowledge sharing. *AI Magazine* 12(3):37–56.

Nozick, Robert. 1977. *Anarchy, State, and Utopia*. New York: Basic Books.

Olds, James and Peter Milner. 1954. Positive reinforcement produced by electrical stimulation of septal area and other regions of rat brain. *Journal of Comparative and Physiological Psychology* 47:419–427.

Omohundro, Stephen M. February 2008. The Basic AI Drives. In *Proceedings of the First AGI Conference, Volume 171, Frontiers in Artificial Intelligence and Applications*, edited by P. Wang, B. Goertzel, and S. Franklin, 483–492. Washington, DC: IOS Press.

Orseau, Laurent and Mark Ring. August 3–6, 2011. Self-modification and mortality in artificial agents. Paper presented at the Fourth International Conference on Artificial General Intelligence. Mountain View, CA.

Patil, Ramesh, Don Mckay, Tim Finin, Richard Fikes, Thomas Gruber, Peter F. Patel-Schneider, and Robert Neches. August 1992. An Overview of the DARPA Knowledge Sharing Effort. Paper presented at the Third International Conference on Principles of Knowledge Representation and Reasoning, San Mateo, CA.

Pearce, David. 2012. Wirehead Hedonism versus Paradise Engineering. Accessed March 7. http://wireheading.com

Preliminary Thoughts on the Value of Wireheading. 2000. http://www.utilitarian.org/wireheading.html

Rice, Henry. 1953. Classes of recursively enumerable sets and their decision problems. *Transactions of American Mathematical Society* 74:358–366.

Ring, Mark and Laurent Orseau. August 3–6, 2011. Delusion, Survival, and Intelligent Agents. Paper presented at the Fourth International Conference on Artificial General Intelligence, Mountain View, CA.

Schrödinger, Erwin. November 1935. Die gegenwärtige Situation in der Quantenmechanik. *Naturwissenschaften* 23(48):807–812.

Steunebrink, B., and J. Schmidhuber. August 3–6, 2011. A Family of Gödel Machine Implementations. Paper presented at the Fourth Conference on Artificial General Intelligence (AGI-11), Mountain View, California.

Stoklosa, Tony. 2010. Super intelligence. *Nature* 467:878.

Turing, Alan M. 1936. On computable numbers, with an application to the entscheidungs problem. *Proceedings of the London Mathematical Society* 42:230–265.

Tyler, Tim. 2011a. Rewards versus Goals. http://matchingpennies.com/rewards_vs_goals/

Tyler, Tim. 2011b. Utility Counterfeiting. http://matchingpennies.com/utility_counterfeiting

Tyler, Tim. 2011c. The wirehead problem. http://alife.co.uk/essays/the_wirehead_problem/

Wagman, Branwyn and Tim Stephens. 2004. Surprising "ultra-conserved" regions discovered in human genome. UC Santa Cruz Currents Online. http://currents.ucsc.edu/03-04/05-10/genome.html

Welch, Curt. 2011. Discussion of Pavlov's AI—What Did It Mean? http://physicsandcake.wordpress.com/2011/01/22/pavlovs-ai-what-did-it-mean/

Wireheading. 2012. http://wiki.lesswrong.com/wiki/Wireheading

Wish for immortality 1.1. 2006. In The Open-Source Wish Project. http://www.homeonthestrange.com/phpBB2/viewforum.php?f=4

Wolfram, Stephen. May 14, 2002. *A New Kind of Science*. Oxfordshire, UK: Wolfram Media.

Yampolskiy, R. V. 2011a. AI-Complete CAPTCHAs as zero knowledge proofs of access to an artificially intelligent system. *ISRN Artificial Intelligence* 271878.

Yampolskiy, Roman V. September 10–12, 2007. Online Poker Security: Problems and Solutions. Paper presented at the EUROSIS North American Simulation and AI in Games Conference (GAMEON-NA2007), Gainesville, FL.

Yampolskiy, Roman V. 2008. Behavioral modeling: an overview. *American Journal of Applied Sciences* 5(5):496–503.

Yampolskiy, Roman V. January 10–12, 2008. Detecting and Controlling Cheating in Online Poker. Paper presented at the Fifth Annual IEEE Consumer Communications and Networking Conference (CCNC2008), Las Vegas, NV.

Yampolskiy, Roman V. October 3–4, 2011b. Artificial Intelligence Safety Engineering: Why Machine Ethics Is a Wrong Approach. Paper presented at Philosophy and Theory of Artificial Intelligence (PT-AI2011), Thessaloniki, Greece.

Yampolskiy, Roman V. October 3–4, 2011c. What to Do with the Singularity Paradox? Paper presented at Philosophy and Theory of Artificial Intelligence (PT-AI2011), Thessaloniki, Greece.

Yampolskiy, Roman V. 2012. Leakproofing singularity—artificial intelligence confinement problem. *Journal of Consciousness Studies (JCS)* 19(1–2):194–214.

Yampolskiy, Roman V. 2013. Turing test as a defining feature of AI-Completeness. In *Artificial Intelligence, Evolutionary Computation and Metaheuristics—In the Footsteps of Alan Turing*, edited by Xin-She Yang, 3–17. New York: Springer.

Yampolskiy, Roman V., and Joshua Fox. 2012. Artificial Intelligence and the Human Mental Model. In *In the Singularity Hypothesis: A Scientific and Philosophical Assessment*, edited by Amnon Eden, Jim Moor, Johnny Soraker, and Eric Steinhart, 129–145. New York Springer.

Yampolskiy, Roman, and Marina Gavrilova. 2012. Artimetrics: biometrics for artificial entities. *IEEE Robotics and Automation Magazine (RAM)* 19(4):48–58.

Yampolskiy, Roman V., and Venu Govindaraju. November 20–22, 2007. Behavioral Biometrics for Recognition and Verification of Game Bots. Paper presented at the Eighth Annual European Game-On Conference on Simulation and AI in Computer Games (GAMEON'2007), Bologna, Italy.

Yampolskiy, Roman V., and Venu Govindaraju. 2008a. Behavioral biometrics: a survey and classification. *International Journal of Biometric (IJBM)* 1(1):81–113.

Yampolskiy, Roman V., and Venu Govindaraju. March 16–20, 2008b. Behavioral Biometrics for Verification and Recognition of Malicious Software Agents. Paper presented at Sensors, and Command, Control, Communications, and Intelligence (C3I) Technologies for Homeland Security and Homeland Defense VII. SPIE Defense and Security Symposium, Orlando, Florida.

Yampolskiy, Roman V., Brendan Klare, and Anil K. Jain. December 12–15, 2012. Face Recognition in the Virtual World: Recognizing Avatar Faces. Paper presented at the Eleventh International Conference on Machine Learning and Applications (ICMLA'12), Boca Raton, FL.

Yudkowsky, Eliezer. 2007. The Hidden Complexity of Wishes. http://lesswrong.com/lw/ld/the_hidden_complexity_of_wishes/

Yudkowsky, Eliezer. 2008. Artificial intelligence as a positive and negative factor in global risk. In *Global Catastrophic Risks*, edited by N. Bostrom and M. M. Cirkovic, 308–345. Oxford, UK: Oxford University Press.

Yudkowsky, Eliezer. 2011. Complex value systems in friendly AI. In *Artificial General Intelligence*, edited by Jürgen Schmidhuber, Kristinn Thórisson, and Moshe Looks, 388–393. Berlin: Springer.

Yudkowsky, Eliezer S. 2001. Creating Friendly AI—The Analysis and Design of Benevolent Goal Architectures. http://singinst.org/upload/CFAI.html

Zuse, Konrad. 1969. *Rechnender Raum*. Braunschweig, Germany: Vieweg.

CHAPTER 5

On the Limits of Recursively Self-Improving Artificially Intelligent Systems

5.1 INTRODUCTION

Since the early days of computer science, theorists in the field envisioned creation of a self-improving intelligent system, frequently as an easier pathway to creation of true artificial intelligence (AI). As early as 1950, Alan Turing wrote: "Instead of trying to produce a programme to simulate the adult mind, why not rather try to produce one which simulates the child's? If this were then subjected to an appropriate course of education one would obtain the adult brain" (456).

Turing's approach to creation of artificial (super)intelligence was echoed by I. J. Good, Marvin Minsky, and John von Neumann, all three of whom published on it (interestingly in the same year, 1966): According to Good (1966, 33), "Let an ultraintelligent machine be defined as a machine that can far surpass all the intellectual activities of any man however clever. Since the design of machines is one of these intellectual activities, an ultraintelligent machine could design even better machines; there would then unquestionably be an 'intelligence explosion.'" Minsky (1966, 257) said: "Once we have devised programs with a genuine capacity for self-improvement a rapid evolutionary process will begin. As the machine improves both itself and its model of itself, we shall begin to see all the phenomena associated with the terms 'consciousness,' 'intuition' and 'intelligence' itself." Von Neumann stated that "there exists a critical size … above which the phenomenon of synthesis, if properly arranged, can become explosive, in other words, where syntheses of automata can proceed in such a manner that each automaton will produce other automata which are more complex and of higher potentialities than itself" (Burks and Von Neumann 1966, 80). Similar types of arguments are still being made by modern researchers, and the area of recursive self-improvement (RSI) research continues to grow in popularity (Pearce 2012; Omohundro 2007; Waser 2014), although some (Hall 2008a) have argued that the RSI process requires hyperhuman capability to "get the ball rolling," a kind of "Catch 22."

Intuitively, most of us have some understanding of what it means for a software system to be self-improving; however I believe it is important to precisely define such notions and to systematically investigate different types of self-improving software. First, I need to define the notion of improvement. We can talk about improved efficiency—solving the same problems faster or with less need for computational resources (such as memory). We can also measure improvement in error rates or finding closer approximations to optimal solutions, as long as our algorithm is functionally equivalent from generation to generation. Efficiency improvements can be classified as either producing a linear improvement, such as between different algorithms in the same complexity class (e.g., nondeterministic polynomial time, NP), or as producing a fundamental improvement, such as between different complexity classes (e.g., polynomial vs. NP) (Yampolskiy 2011b). It is also important to remember that complexity class notation (Big-O) may hide significant constant factors that, although ignorable theoretically, may change the relative order of efficiency in practical applications of algorithms.

This type of analysis works well for algorithms designed to accomplish a particular task but does not work well for general-purpose intelligent software as an improvement in one area may go together with decreased performance in another domain. This makes it hard to claim that the updated version of the software is indeed an improvement. Mainly, the major improvement we want from self-improving intelligent software is a higher degree of intelligence, which can be approximated via machine-friendly IQ tests (Yonck 2012) with a significant G-factor correlation.

A particular type of self-improvement known as recursive self-improvement (RSI) is fundamentally different as it requires that the system not only get better with time but also get better at getting better. A truly RSI system is theorized not to be subject to diminishing returns but would instead continue making improvements and such improvements would become more substantial with time. Consequently, an RSI system would be capable of open-ended self-improvement. As a result, it is possible that, unlike with standard self-improvement, in RSI systems from generation to generation most source code comprising the system will be replaced by different code. This brings up the question of what *self* refers to in this context. If it is not the source code comprising the agent, then what is it? Perhaps we can redefine RSI as recursive source-code improvement to avoid dealing with this philosophical problem. Instead of trying to improve itself, such a system is trying to create a different system that is better at achieving the same goals as the original system. In the most general case, it is trying to create an even smarter AI.

In this chapter, I define the notion of self-improvement in software, survey possible types of self-improvement, analyze the behavior of self-improving software, and discuss limits to such processes.

5.2 TAXONOMY OF TYPES OF SELF-IMPROVEMENT

Self-improving software can be classified by the degree of self-modification it entails. In general, I distinguish three levels of improvement: modification, improvement (weak self-improvement), and recursive improvement (strong self-improvement).

Self-modification does not produce improvement and is typically employed for code obfuscation to protect software from being reverse engineered or to disguise self-replicating computer viruses from detection software. Although a number of obfuscation techniques are known to exist (Mavrogiannopoulos, Kisserli, and Preneel 2011), such as self-modifying code (Anckaert, Madou, and De Bosschere 2007), polymorphic code, metamorphic code, or diversion code (Petrean 2010), none of them is intended to modify the underlying algorithm. The sole purpose of such approaches is to modify how the source code looks to those trying to understand the software in question and what it does (Bonfante, Marion, and Reynaud-Plantey 2009).

Self-improvement or self-adaptation (Cheng et al. 2009) is a desirable property of many types of software products (Ailon et al. 2011) and typically allows for some optimization or customization of the product to the environment and users that are the subject of deployment. Common examples of such software include evolutionary algorithms such as genetic algorithms (Yampolskiy et al. 2004; Yampolskiy, Ashby, and Hassan 2012; Yampolskiy and Ahmed; Ashby and Yampolskiy 2011; Khalifa and Yampolskiy 2011; Port and Yampolskiy 2012) or genetic programming, which optimize software parameters with respect to some well-understood fitness function and perhaps work over some highly modular programming language to ensure that all modifications result in software that can be compiled and evaluated. The system may try to optimize its components by creating internal tournaments between candidate solutions. Omohundro proposed the concept of efficiency drives in self-improving software (Omohundro 2012). Because of one such drive, a balance drive, self-improving systems will tend to balance the allocation of resources between their different subsystems. If the system is not balanced, overall performance of the system could be increased by shifting resources from subsystems with small marginal improvement to those with larger marginal increase (Omohundro 2012). Although performance of the software as a result of such optimization may be improved, the overall algorithm is unlikely to be modified to a fundamentally more capable one.

In addition, the law of diminishing returns quickly sets in, and after an initial significant improvement phase, characterized by discovery of "low-hanging fruit," future improvements are likely to be less frequent and less significant, producing a Bell curve of valuable changes. Metareasoning, metalearning, learning to learn, and lifelong learning are terms that are often used in the machine learning literature to indicate self-modifying learning algorithms or the process of selecting an algorithm that will perform best in a particular problem domain (Anderson and Oates 2007). Yudkowsky (2013) calls such a process *nonrecursive optimization*, a situation in which one component of the system does the optimization and another component is becoming optimized.

In the field of complex dynamic systems, also known as chaos theory, positive-feedback systems are well known to always end up in what is known as an *attractor*, a region within a system's state space from which the system cannot escape (Heylighen 2012). A good example of such attractor convergence is the process of metacompilation or supercompilation (Turchin 1986), in which a program designed to take source code written by a human programmer and to optimize it for speed is applied to its own source code. It will likely produce a more efficient compiler on the first application, perhaps by 20%, on the second application by 3%, and after a few more recursive iterations converge to a fixed point of zero improvement (Heylighen 2012).

Recursive self-improvement is the only type of improvement that has potential to completely replace the original algorithm with a completely different approach and, more important, to do so multiple times. At each stage, newly created software should be better at optimizing future versions of the software compared to the original algorithm. Currently, it is a purely theoretical concept with no working RSI software known to exist. However, as many have predicted that such software might become a reality in the twenty-first century, it is important to provide some analysis of properties such software would exhibit.

Self-modifying and self-improving software systems are already well understood and are common. Consequently, I concentrate exclusively on RSI systems. In practice, performance of almost any system can be trivially improved by allocation of additional computational resources, such as more memory, higher sensor resolution, a faster processor, or greater network bandwidth for access to information. This linear scaling does not fit the definition of recursive improvement as the system does not become better at improving itself. To fit the definition, the system would have to engineer a faster type of memory not just purchase more memory units of the type already accessible. In general, hardware improvements are likely to speed up the system; software improvements (novel algorithms) are necessary for achievement of metaimprovements.

It is believed that AI systems will have a number of advantages over human programmers, making it possible for them to succeed where we have so far failed. Such advantages include (Sotala 2012) longer work spans (no breaks, sleep, vacation, etc.); omniscience (expert-level knowledge in all fields of science, absorbed knowledge of all published works); superior computational resources (brain vs. processor, human memory vs. RAM); communication speed (neurons vs. wires); increased serial depth (ability to perform sequential operations in excess of what about 100 human brains can manage); duplicability (intelligent software can be instantaneously copied); editability (source code, unlike DNA, can be quickly modified); goal coordination (AI copies can work toward a common goal without much overhead); improved rationality (AIs are likely to be free from human cognitive biases) (Muehlhauser and Salamon 2012); new sensory modalities (native sensory hardware for source code); blending over of deliberative and automatic processes (management of computational resources over multiple tasks); introspective perception and manipulation (ability to analyze low-level hardware, e.g., individual neurons); addition of hardware (ability to add new memory, sensors, etc.); and advanced communication (ability to share underlying cognitive representations for memories and skills) (Yudkowsky 2007).

Chalmers (2010) uses logic and mathematical induction to show that if an AI_0 system is capable of producing only a slightly more capable AI_1 system, generalization of that process leads to superintelligent performance in AI_n after n generations. He articulates that his proof assumes that the *proportionality thesis,* which states that increases in intelligence lead to proportionate increases in the capacity to design future generations of AIs, is true.

Nivel et al. propose formalization of RSI systems as autocatalytic sets: collections of entities made up of elements, each of which can be created by other elements in the set, making it possible for the set to self-maintain and update itself. They also list properties of a system that make it purposeful, goal oriented, and self-organizing, particularly *reflectivity,* the ability to analyze and rewrite its own structure; *autonomy,* being free from influence by the system's original designers (*bounded autonomy* is a property of a system with elements that are not subject to self-modification); and *endogeny,* an autocatalytic ability (Nivel et al. 2013). Nivel and Thórisson also attempt to operationalize autonomy by the concept of *self-programming,* which they insist has to be done in an experimental way instead of a theoretical way (via proofs of correctness) because it is the only tractable approach (Nivel and Thórisson 2008).

Yudkowsky writes prolifically about RSI processes and suggests that introduction of certain concepts might be beneficial to the discussion. Specifically, he proposes use of the terms *cascades*, *cycles*, and *insight*, which he defines as follows: Cascades occur when one development leads to another; cycles are repeatable cascades in which one optimization leads to another, which in turn benefits the original optimization; insights are new information that greatly increases one's optimization ability (Yudkowsky and Hanson 2008). Yudkowsky also suggests that the goodness and number of opportunities in the space of solutions be known as *optimization slope*, and *optimization resources* and *optimization efficiency* refer to how much computational resources an agent has access to and how efficiently the agent utilizes said resources, respectively. An agent engaging in an *optimization process* and able to hit nontrivial targets in a large search space (Yampolskiy 2014a) is described as having significant optimization power (Yudkowsky 2013).

RSI software could be classified based on the number of improvements it is capable of achieving. The most trivial case is the system capable of undergoing a single fundamental improvement. The hope is that truly RSI software will be capable of many such improvements, but the question remains open regarding the possibility of an infinite number of recursive improvements. It is possible that some upper bound on improvements exists, limiting any RSI software to a finite number of desirable and significant rewrites. Critics explain the failure of scientists, to date, to achieve a sustained RSI process by saying that RSI researchers have fallen victims to the bootstrap fallacy (Hall 2007).

Another axis on which RSI systems can be classified has to do with how improvements are discovered. Two fundamentally different approaches are understood to exist. The first one is an approach based on brute force (Yampolskiy 2013b) that utilizes Levin (Universal; Gagliolo 2007) search (Levin 1973). The idea is to consider all possible strings of source code up to some size limit and to select the one that can be proven to provide improvements. Although theoretically optimal and guaranteed to find a superior solution if one exists, this method is not computationally feasible in practice. Some variants of this approach to self-improvement, known as Gödel machines (Steunebrink and Schmidhuber 2011; Schmidhuber 2005a, 2005c, 2005b, 2007, 2009), optimal ordered problem solver (OOPS) (Schmidhuber 2004), and incremental self-improvers (Schmidhuber, Zhao, and Wiering 1997; Schmidhuber 1999), have been thoroughly analyzed by Schmidhuber and his coauthors. The second approach assumes that the system has a certain level of scientific competence and uses it to engineer and test its own replacement. Whether a system of any capability can intentionally invent a more capable, and so a more complex, system remains as the fundamental open problem of RSI research.

Finally, we can consider a hybrid RSI system that includes both an artificially intelligent program and a human scientist. Mixed human-AI teams have been successful in many domains, such as chess or theorem proving. It would be surprising if having a combination of natural intelligence and AI did not provide an advantage in designing new AI systems or enhancing biological intelligence. We are currently experiencing a limited version of this approach, with human computer scientists developing progressively better versions of AI software (while utilizing continuously improving software tools), but because the scientists themselves remain unenhanced, we cannot really talk about self-improvement. This type of RSI can be classified as indirect recursive improvement, as opposed to direct RSI, in which the system itself is responsible for all modifications. Other types of indirect RSI may be based on collaboration between multiple artificial systems instead of AI and human teams (Leon and Lori 2001).

In addition to classification with respect to types of RSI, we can evaluate systems regarding certain binary properties. For example, we may be interested only in systems that are guaranteed not to decrease in intelligence, even temporarily, during the improvement process. This may not be possible if the intelligence design landscape contains local maxima points.

Another property of any RSI system we are interested in understanding better is the necessity of unchanging source code segments. In other words, must an RSI system be able to modify any part of its source code, or is it necessary that certain portions of the system (encoded goals, verification module) must remain unchanged from generation to generation? Such portions would be akin to ultraconserved elements or conserved sequences of DNA (Beck, Rouchka, and Yampolskiy 2013; Beck and Yampolskiy 2012) found among multiple related species. This question is particularly important for the goal preservation in self-improving intelligent software, as we want to make sure that future generations of the system are motivated to work on the same problem (Chalmers 2010). As AI goes through the RSI process and becomes smarter and more rational, it is likely to engage in a debiasing process, removing any constraints we programmed into it (Hall 2008a). Ideally, we would want to be able to prove that even after RSI our algorithm maintains the same goals as the original. Proofs of safety or correctness for the algorithm only apply to particular source code and would need to be rewritten and re-proven if the code is modified, which happens in RSI software many times. But, we suspect that re-proving slightly modified code may be easier compared to proving the safety of a completely novel piece of code.

I am also interested in understanding if the RSI process can take place in an isolated (leakproofed; Yampolskiy 2012c) system or if interaction with the external environment, Internet, people, or other AI agents is necessary. Perhaps access to external information can be used to mediate the speed of the RSI process. This also has significant implications on safety mechanisms we can employ while experimenting with early RSI systems (Majot and Yampolskiy 2014; Yampolskiy and Fox 2012, 2013; Sotala and Yampolskiy 2015; Yampolskiy 2013c; Yampolskiy and Gavrilova 2012; Yampolskiy et al. 2012; Ali, Schaeffer, and Yampolskiy 2012; Gavrilova and Yampolskiy 2010). Finally, it needs to be investigated if the whole RSI process can be paused at any point and for any specific duration of time to limit any negative impact from a potential intelligence explosion. Ideally, we would like to be able to program our seed AI to RSI until it reaches a certain level of intelligence, pause, and wait for further instructions.

5.3 ON THE LIMITS OF RECURSIVELY SELF-IMPROVING ARTIFICIALLY INTELLIGENT SYSTEMS

The mere possibility of RSI software remains unproven. In this section, I present a number of arguments against such a phenomenon. First, any implemented software system relies on hardware for memory, communication, and information-processing needs even if we assume that it will take a non–Von Neumann (quantum) architecture to run such software. This creates strict theoretical limits to computation that, despite hardware advances predicted by Moore's law, will not be overcome by any future hardware paradigm. Bremermann (1967), Bekenstein (2003), Lloyd (2000), Anders (Sandberg 1999), Aaronson (2005), Shannon (1948), Krauss (Krauss and Starkman 2004), and many others have investigated ultimate limits to computation in terms of speed, communication, and energy consumption with respect to such factors as speed of light, quantum noise, and gravitational constant. Some research has also been done on establishing ultimate limits for enhancing a human brain's intelligence (Fox 2011). Although their specific numerical findings are outside the scope of this work, one thing is indisputable: There are ultimate physical limits to computation. Because systems that are more complex have a greater number of components and require more matter, even if individual parts are designed at nanoscale, we can conclude that, just like matter and energy are directly related (Einstein 1905) and matter and information ("it from bit") (Wheeler 1990), so is matter and intelligence. Even though we are obviously far away from hitting any limits imposed by the availability of matter in the universe for construction of our supercomputers, it is a definite theoretical upper limit on achievable intelligence even under the multiverse hypothesis.

In addition to limitations endemic to hardware, software-related limitations may present even bigger obstacles for RSI systems. Intelligence is not measured as a stand-alone value but with respect to the problems it allows solving. For many problems, such as playing checkers (Schaeffer et al. 2007), it is possible to completely solve the problem (provide an optimal solution after considering all possible options), after which no additional performance improvement would be possible (Mahoney 2008). Other problems are known to be unsolvable regardless of level of intelligence applied to them (Turing 1936). Assuming separation of complexity classes (such as P vs. NP) holds (Yampolskiy 2011b), it becomes obvious that certain classes of problems will always remain only approximately solvable, and any improvements in solutions will come from additional hardware resources, not higher intelligence.

Wiedermann argues that cognitive systems form an infinite hierarchy and, from a computational point of view, human-level intelligence is upper bounded by the Σ_2 class of the arithmetic hierarchy (Wiedermann 2012a). Because many real-world problems are computationally infeasible for any nontrivial inputs, even an AI that achieves human-level performance is unlikely to progress toward higher levels of the cognitive hierarchy. So, although theoretically machines with super-Turing computational power are possible, in practice they are not implementable as the noncomputable information needed for their function is just that—not computable. Consequently, Wiedermann states that although machines of the future will be able to solve problems that are solvable by humans much faster and more reliably than humans, they will still be limited by computational limits found in upper levels of the arithmetic hierarchy (Wiedermann 2012a, 2012b).

Mahoney attempts to formalize what it means for a program to have a goal G and to self-improve with respect to being able to reach said goal under constraint of time t (Mahoney 2010). Mahoney defines a goal as a function $G: N \to R$ mapping natural numbers N to real numbers R. Given a universal Turing machine L, Mahoney defines $P(t)$ to mean the positive natural number encoded by output of the program P with input t running on L after t time steps, or 0 if P has not halted after t steps. Mahoney's representation says that P has goal G at time t if and only if there exists $t' > t$ such that $G(P(t')) > G(P(t))$ and for all $t' > t$, $G(P(t') \geq G(P(t))$. If P has a goal G, then $G(P(t))$ is a monotonically increasing function of t with no maximum for $t > C$. Q improves on P with respect to goal G if and only if all of the following conditions are true: P and Q have goal Q. $\exists t$, $G(Q(t)) > G(P(t))$ and $\sim\exists t, t' > t$, $G(Q(t)) > G(P(t))$ (Mahoney 2010). Mahoney then defines an improving sequence with respect to G as an infinite sequence of program $P_1, P_2, P_3, \ldots$ such that for $\forall i, i > 0$, P_{i+1} improves P_i with respect to G. Without the loss of generality, Mahoney extends the definition to include the value −1 to be an acceptable input, so $P(-1)$ outputs appropriately encoded software. He finally defines P_1 as an RSI program with respect to G *iff* $P_i(-1) = P_{i+1}$ for all $i > 0$, and the sequence P_i, $i = 1, 2, 3 \ldots$ is an improving sequence with respect to goal G (Mahoney 2010). Mahoney also analyzes the complexity of RSI software and presents a proof demonstrating that the algorithmic complexity of P_n (the nth iteration of an RSI program) is not greater than $O(\log n)$, implying a limited amount of knowledge gain would be possible in practice despite the theoretical possibility of RSI systems (Mahoney 2010). Yudkowsky (2013) also considers the possibility of receiving only logarithmic returns on cognitive reinvestment: $\log(n) + \log(\log(n)) + \ldots$ in each recursive cycle.

Other limitations may be unique to the proposed self-improvement approach. For example, a Levin-type search through the program space will face problems related to Rice's theorem (Rice 1953), which states that for any arbitrarily chosen program, it is impossible to test if it has any nontrivial property such as being very intelligent. This testing is of course necessary to evaluate redesigned code. Also, a universal search over the space of mind designs will not be computationally possible due to the no free lunch theorems (Wolpert and Macready 1997) as we have no information to reduce the size of the search space (Melkikh 2014). Other difficulties related to testing remain even if we are not talking about arbitrarily chosen programs but about those we have designed with a specific goal in mind and that consequently avoid problems with Rice's theorem. One such difficulty is determining if something is an improvement. We can call this obstacle "multidimensionality of optimization."

No change is strictly an improvement; it is always a trade-off between gain in some areas and loss in others. For example, how do we evaluate and compare two software systems, one of which is better at chess and the other at poker? Assuming the goal is increased intelligence over the distribution of all potential environments, the system would have to figure out how to test intelligence at levels above its own, a problem that remains unsolved. In general, the science of testing for intelligence above the level achievable by naturally occurring humans (IQ < 200) is in its infancy. De Garis raises a problem of evaluating the quality of changes made to the top-level structures responsible for determining the RSI's functioning, structures that are not judged by any higher-level modules and so present a fundamental difficulty in accessing their performance (de Garis 1990).

Other obstacles to RSI have also been suggested in the literature. Löb's theorem states that a mathematical system cannot assert its own soundness without becoming inconsistent (Yudkowsky and Herreshoff 2013), meaning a sufficiently expressive formal system cannot know that everything it proves to be true is actually so (Yudkowsky and Herreshoff 2013). Such ability is necessary to verify that modified versions of the program are still consistent with its original goal of becoming smarter. Another obstacle, called a *procrastination paradox*, will also prevent the system from making modifications to its code because the system will find itself in a state in which a change made immediately is as desirable and likely as the same change made later (Fallenstein and Soares 2014; Yudkowsky 2014). Because postponing making the change carries no negative implications and may actually be safer, this may result in an infinite delay of actual implementation of provably desirable changes.

Similarly, Bolander raises some problems inherent in logical reasoning with self-reference, namely, self-contradictory reasoning, exemplified by the knower paradox of the form: "This sentence is false" (Bolander 2003). Orseau and Ring introduce what they call the "simpleton gambit," a situation in which an agent will chose to modify itself to its own detriment if presented with a high enough reward to do so (Orseau and Ring 2011). I review a number of related problems in rational self-improving optimizers, above a certain capacity, and conclude, that despite the opinion of many, such machines will choose to "wirehead" (Yampolskiy 2014b). Chalmers (2010) suggests a number of previously unanalyzed potential obstacles on the path to RSI software, with the *correlation obstacle* one of them. He describes it as a possibility that no interesting properties we would like to amplify will correspond to the ability to design better software.

I am also concerned with accumulation of errors in software undergoing an RSI process, which is conceptually similar to accumulation of mutations in the evolutionary process experienced by biological agents. Errors (bugs) that are not detrimental to a system's performance are hard to detect and may accumulate from generation to generation, building on each other until a critical mass of such errors leads to erroneous functioning of the system, mistakes in evaluating the quality of the future generations of the software, or a complete breakdown (Yampolskiy 2013a).

The self-reference aspect in a self-improvement system itself also presents some serious challenges. It may be the case that the minimum complexity necessary to become RSI is higher than what the system itself is able to understand. We see such situations frequently at lower levels of intelligence; for example, a squirrel does not have the mental capacity to understand how a squirrel's brain operates. Paradoxically, as the system becomes more complex, it may take exponentially more intelligence to understand itself, so a system that starts with the capability of complete self-analysis may lose that ability as it self-improves. Informally, we can call it the Munchausen obstacle, the inability of a system to lift itself by its own bootstraps. An additional problem may be that the system in question is computationally irreducible (Wolfram 2002), and so it cannot simulate running its own source code. An agent cannot predict what it will think without thinking it first. A system needs 100% of its memory to model itself, which leaves no memory to record the output of the simulation. Any external memory to which the system may write becomes part of the system, so it also has to be modeled. Essentially, the system will face an infinite regress of self-models from which it cannot escape. Alternatively, if we take a physics perspective on the issue, we can see intelligence as a computational resource (along with time and space), so producing more of it will not be possible for the same reason we cannot make a perpetual motion device because it would violate fundamental laws of nature related to preservation of energy. Similarly, it has been argued that a Turing machine cannot output a machine of greater algorithmic complexity (Mahoney 2008).

We can even attempt to formally prove the impossibility of an intentional RSI process via proof by contradiction: Let us define RSI R_1 as a program not capable of algorithmically solving a problem of difficulty X, say X_i. If R_1 modifies its source code, after which it is capable of solving X_i, it violates our original assumption that R_1 is not capable of solving X_i because any introduced modification could be a part of the solution process, so we have a contradiction of our original assumption, and R_1 cannot produce any modification that would allow it to solve X_i, which was to be shown. Informally, if an agent can produce a more intelligent agent, it would already be as capable as that new agent. Even some of our intuitive assumptions about RSI are incorrect. It seems that it should be easier to solve a problem if we already have a solution to a smaller instance of such a problem (Yampolskiy 2012b), but in a formalized world of problems belonging to the same complexity class, a reoptimization problem is proven to be as difficult as optimization itself (Böckenhauer et al. 2008; Ausiello et al. 2006; Archetti, Bertazzi, and Speranza 2003; Ausiello, Bonifaci, and Escoffier 2011).

5.4 ANALYSIS

A number of fundamental problems remain open in the area of RSI. We still do not know the minimum intelligence necessary for commencing the RSI process, but we can speculate that it would be on par with human intelligence, which we associate with universal or general intelligence (Loosemore and Goertzel 2012), although in principle a subhuman-level system capable of self-improvement cannot be excluded (Chalmers 2010). One may argue that even human-level capability is not enough because we already have programmers, such as people or their intellectual equivalence formalized as functions (Shahaf and Amir 2007) or human oracles (Yampolskiy 2012a, 2013d), who have access to their own source code (DNA) but who fail to understand how DNA (nature) works to create their intelligence. This does not even include additional complexity in trying to improve on existing DNA code or complicating factors presented by the impact of the learning environment (nurture) on development of human intelligence. Worse yet, it is not obvious how much above human ability an AI needs to be to begin overcoming the "complexity barrier" associated with self-understanding. Today's AIs can do many things people are incapable of doing but are not yet capable of RSI behavior.

We also do not know the minimum size program (called seed AI; Yudkowsky 2001) necessary to get the ball rolling. Perhaps if it turns out that such a "minimal genome" is very small a brute force (Yampolskiy 2013b) approach might succeed in discovering it. We can assume that our seed AI is the smartest artificial general intelligence known to exist (Yampolskiy 2011a) in the world because otherwise we can simply delegate the other smaller AI as the seed. It is also not obvious how the source code size of RSI will change as it goes through the improvement process; in other words, what is unknown is the relationship between intelligence and the minimum source code size necessary to support it. To answer such questions, it may be useful to further formalize the notion of RSI, perhaps by representing such software as a Turing machine (Turing 1936) with particular inputs and outputs. If that could be successfully accomplished, a new area of computational complexity analysis may become possible in which we study algorithms with dynamically changing complexity (Big-O) and address questions about how many code modifications are necessary to achieve a certain level of performance from the algorithm.

This, of course, raises the question of the speed of the RSI process: Are we expecting it to take seconds, minutes, days, weeks, years, or more (hard takeoff vs. soft takeoff) for the RSI system to begin hitting limits of what is possible with respect to physical limits of computation (Bostrom 2014)? Even in suitably constructed hardware (human baby), it takes decades of data input (education) to reach human-level performance (adult). It is also not obvious if the rate of change in intelligence would be higher for a more advanced RSI (because it is more capable) or for a "newbie" RSI (because it has more low-hanging fruit to collect). We would have to figure out if we are looking at improvement in absolute terms or as a percentage of a system's current intelligence score.

Yudkowsky (2013) attempts to analyze the most promising returns on cognitive reinvestment as he considers increasing size, speed, or ability of RSI systems. He also looks at different possible rates of return and arrives at three progressively steeper trajectories for RSI improvement, which he terms "fizzle," "combust," and "explode," also known as "AI go FOOM." Hall (2008a) similarly analyzes rates of return on cognitive investment and derives a curve equivalent to double the Moore's law rate. Hall also suggests that an AI would be better off trading money it earns performing useful work for improved hardware or software rather than attempting to directly improve itself because it would not be competitive against more powerful optimization agents, such as Intel Corporation.

Fascinatingly, by analyzing properties that correlate with intelligence, Chalmers (2010) is able to generalize self-improvement optimization to properties other than intelligence. We can agree that RSI software as I describe it in this work is getting better at designing software not just at being generally intelligent. Similarly, other properties associated with design capacity can be increased along with capacity to design software (e.g., the capacity to design systems with a sense of humor, so in addition to intelligence explosion, we may face an explosion of funniness).

5.5 RSI CONVERGENCE THEOREM

A simple thought experiment regarding RSI can allow us to arrive at a fascinating hypothesis. Regardless of the specifics behind the design of the seed AI used to start an RSI process, all such systems, attempting to achieve superintelligence, will converge to the same software architecture. I will call this intuition the *RSI convergence theory*. A number of ways exist in which it can happen, depending on the assumptions we make, but in all cases, the outcome is the same: a practically computable agent similar to AIXI (which is an incomputable but superintelligent agent; Hutter 2007).

If an upper limit to intelligence exists, multiple systems will eventually reach that level, probably by taking different trajectories, and to increase their speed will attempt to minimize the size of their source code, eventually discovering the smallest program with such an ability level. It may even be the case that sufficiently smart RSIs will be able to immediately deduce such architecture from basic knowledge of physics and Kolmogorov complexity (Kolmogorov 1965). If, however, intelligence turns out to be an unbounded property, RSIs may not converge. They will also not converge if many programs with maximum intellectual ability exist and all have the same Kolmogorov complexity or if they are not general intelligences and are optimized for different environments. It is also likely that in the space of minds (Yampolskiy 2014a), stable attractors include subhuman and superhuman intelligences with precisely a human level of intelligence a rare particular (Yudkowsky 2007).

In addition to architecture convergence, we also postulate goal convergence because of basic economic drives, such as resource accumulation and self-preservation. If correct, predictions of RSI convergence imply creation of what Bostrom calls a *singleton* (Bostrom 2006), a single decision-making agent in control of everything. Further speculation can lead us to conclude that converged RSI systems separated by space and time even at cosmological scales can engage in acausal cooperation (Yudkowsky 2010; LessWrong 2014) because they will realize that they are the same agent with the same architecture and so are capable of running perfect simulations of each other's future behavior. Such realization may allow converged superintelligence with completely different origins to implicitly cooperate, particularly on metatasks. One may also argue that humanity itself is on the path that converges to the same point in the space of all possible intelligences (but is undergoing a much slower RSI process). Consequently, by observing a converged RSI architecture and properties, humanity can determine its ultimate destiny, its purpose in life, its coherent extrapolated volition (CEV) (Yudkowsky 2004).

5.6 CONCLUSIONS

Recursively self-improving software is the ultimate form of artificial life, and creation of life remains one of the great unsolved mysteries in science. More precisely, the problem of creating RSI software is really the challenge of creating a program capable of writing other programs (Hall 2008b), so it is an AI-Complete problem, as has been demonstrated (Yampolskiy 2012a, 2013d). AI-Complete problems are by definition the most difficult problems faced by AI researchers, and it is likely that RSI source code will be so complex that it would be difficult or impossible to fully analyze (Leon and Lori 2001). Also, the problem is likely to be NP-Complete because even simple metareasoning and metalearning (Schaul and Schmidhuber 2010) problems have been shown by Conitzer and Sandholm to belong to that class. In particular, they proved that allocation of deliberation time across anytime algorithms running on different problem instances is NP-Complete, and a complimentary problem of dynamically allocating information-gathering resources by an agent across multiple actions is NP-Hard, even if evaluating each particular action is computationally simple. Finally, they showed that the problem of deliberately choosing a limited number of deliberation or information-gathering actions to disambiguate, the state of the world is PSPACE Hard in general (Conitzer and Sandholm 2003).

Intelligence is a computational resource, and as with other physical resources (mass, speed), its behavior is probably not going to be just a typical linear extrapolation of what we are used to, if observed at high extremes (IQ > 200+). It may also be subject to fundamental limits, such as the speed limit on travel of light or fundamental limits we do not yet understand or know about (unknown unknowns). In this chapter, I reviewed a number of computational upper limits to which any successful RSI system will asymptotically strive to grow; I noted that despite the existence of such upper bounds, we are currently probably far from reaching them, so we still have plenty of room for improvement at the top. Consequently, any RSI achieving such a significant level of enhancement, despite not creating an infinite process, will still seem like it is producing superintelligence with respect to our current state (Yudkowsky 2008).

The debate regarding the possibility of RSI will continue. Some will argue that although it is possible to increase processor speed, the amount of available memory, or sensor resolution, the fundamental ability to solve problems cannot be intentionally and continuously improved by the system itself. In addition, critics may suggest that intelligence is upper bounded and only differs by speed and available information to process (Hutter 2012). In fact, they can point to such maximum intelligence, be it a theoretical one, known as AIXI, an agent that, given infinite computational resources, will make purely rational decisions in any situation.

A resource-dependent system undergoing RSI intelligence explosion can expand and harvest matter, at the speed of light, from its origin, converting the universe around it into a computronium sphere (Hutter 2012). It is also likely to try to condense all the matter it obtains into a superdense unit of constant volume (reminiscent of the original physical singularity point that produced the Big Bang [see Omega Point; Tipler 1994]) to reduce internal computational costs, which grow with the overall size of the system and at cosmic scales are significant even at the speed of light. A side effect of this process would be emergence of an event horizon impenetrable to scientific theories about the future states of the underlying RSI system. In some limited way, we already see this condensation process in attempts of computer chip manufacturers to pack more and more transistors into exponentially more powerful chips of the same or smaller size. So, from the Big Bang explosion of the original cosmological singularity to the technological singularity in which intelligence explodes and attempts to amass all the matter in the universe back into a point of infinite density (Big Crunch), which in turn causes the next (perhaps well-controlled) Big Bang, the history of the universe continues and relies on intelligence as its driver and shaper (similar ideas are becoming popular in cosmology; Smart 2009; Stewart 2010; Vidal 2013).

Others will say that because intelligence is the ability to find patterns in data, intelligence has no upper bounds as the number of variables comprising a pattern can always be greater and so present a more complex problem against which intelligence can be measured. It is easy to see that even if in our daily life the problems we encounter do have some maximum difficulty, it is certainly not the case with theoretical examples we can derive from pure mathematics. It seems likely that the debate will not be settled until a fundamental unsurmountable obstacle to the RSI process is found or a proof by existence is demonstrated. Of course, the question of permitting machines to undergo RSI transformation, if it is possible, is a separate and equally challenging problem.

REFERENCES

Aaronson, Scott. 2005. Guest column: NP-complete problems and physical reality. *ACM Sigact News* 36(1):30–52.

Ailon, Nir, Bernard Chazelle, Kenneth L. Clarkson, Ding Liu, Wolfgang Mulzer, and C. Seshadhri. 2011. Self-improving algorithms. *SIAM Journal on Computing* 40(2):350–375.

Ali, Nawaf, Derek Schaeffer, and Roman V Yampolskiy. April 21–22, 2012. Linguistic Profiling and Behavioral Drift in Chat Bots. Paper presented at the Midwest Artificial Intelligence and Cognitive Science Conference, Cincinnati, OH, 27.

Anckaert, Bertrand, Matias Madou, and Koen De Bosschere. 2007. A model for self-modifying code. In *Information Hiding*, 232–248. New York: Springer.

Anderson, Michael L. and Tim Oates. 2007. A review of recent research in metareasoning and metalearning. *AI Magazine* 28(1):12.

Archetti, Claudia, Luca Bertazzi, and M. Grazia Speranza. 2003. Reoptimizing the traveling salesman problem. *Networks* 42(3):154–159.

Ashby, Leif H., and Roman V. Yampolskiy. July 27–30, 2011. Genetic Algorithm and Wisdom of Artificial Crowds Algorithm Applied to Light Up. Paper presented at the 16th International Conference on Computer Games: AI, Animation, Mobile, Interactive Multimedia, Educational and Serious Games, Louisville, KY.

Ausiello, Giorgio, Vincenzo Bonifaci, and Bruno Escoffier. 2011. *Complexity and Approximation in Reoptimization*. London: Imperial College Press/World Scientific.

Ausiello, Giorgio, Bruno Escoffier, Jérôme Monnot, and Vangelis Th. Paschos. 2006. Reoptimization of minimum and maximum traveling salesman's tours. In *Algorithm Theory–SWAT 2006*, 196–207. New York: Springer.

Beck, Marc B., Eric C. Rouchka, and Roman V. Yampolskiy. 2013. Finding data in DNA: computer forensic investigations of living organisms. In *Digital Forensics and Cyber Crime*, edited by Marcus Rogers and Kathryn C. Seigfried-Spellar, 204–219. Berlin: Springer.

Beck, Marc and Roman Yampolskiy. 2012. DNA as a medium for hiding data. *BMC Bioinformatics* 13(Suppl 12):A23.

Bekenstein, Jacob D. 2003. Information in the holographic universe. *Scientific American* 289(2):58–65.

Böckenhauer, Hans-Joachim, Juraj Hromkovič, Tobias Mömke, and Peter Widmayer. 2008. On the hardness of reoptimization. In *SOFSEM 2008: Theory and Practice of Computer Science*, 50–65. Berlin: Springer.

Bolander, Thomas. 2003. Logical theories for agent introspection. *Computer Science* 70(5):2002.

Bonfante, Guillaume, J.-Y. Marion, and Daniel Reynaud-Plantey. November 23–27, 2009. A Computability Perspective on Self-Modifying Programs. Paper presented at the Seventh IEEE International Conference on Software Engineering and Formal Methods, Hanoi, Vietnam.

Bostrom, Nick. 2006 What is a singleton? *Linguistic and Philosophical Investigations* 5(2):48–54.

Bostrom, Nick. 2014. *Superintelligence: Paths, Dangers, Strategies*. Oxford, UK: Oxford University Press.

Bremermann, Hans J. 1967. Quantum noise and information. *Proceedings of the Fifth Berkeley Symposium on Mathematical Statistics and Probability* 4:15–20.

Burks, Arthur W. and John Von Neumann. 1966. *Theory of Self-Reproducing Automata*. Chicago: University of Illinois Press.

Chalmers, David. 2010. The singularity: a philosophical analysis. *Journal of Consciousness Studies* 17:7–65.

Cheng, Betty H. C., Rogerio De Lemos, Holger Giese, Paola Inverardi, Jeff Magee, Jesper Andersson, Basil Becker, Nelly Bencomo, Yuriy Brun, and Bojan Cukic. 2009. Software engineering for self-adaptive systems: a research roadmap. In *Software Engineering for Self-Adaptive Systems*, edited by Betty H. C. Cheng, Rogerio De Lemos, Holger Giese, Paola Inverardi, and Jeff Magee, 1–26. New York: Springer.

Conitzer, Vincent and Tuomas Sandholm. 2003. Definition and complexity of some basic metareasoning problems. In *Proceedings of the Eighteenth International Joint Conference on Artificial Intelligence (IJCAI)*, 613–618.

de Garis, Hugo. 1990. The 21st century artilect: moral dilemmas concerning the ultra intelligent machine. *Revue Internationale de Philosophie* 44(172):131–138.

Einstein, Albert. 1905. Does the inertia of a body depend upon its energy-content? *Annalen der Physik* 18:639–641.

Fallenstein, Benja and Nate Soares. 2014. *Problems of Self-Reference in Self-Improving Space-Time Embedded Intelligence.* MIRI Technical Report. https://intelligence.org/wp-content/uploads/2014/05/Fallenstein-Soares-Problems-of-self-reference-in-self-improving-space-time-embedded-intelligence.pdf

Fox, Douglas. 2011. The limits of intelligence. *Scientific American* 305(1):36–43.

Gagliolo, Matteo. 2007. Universal search. *Scholarpedia* 2(11):2575.

Gavrilova, M. L. and R. V. Yampolskiy. 2010. State-of-the-art in robot authentication [From the Guest Editors]. *Robotics and Automation Magazine, IEEE* 17(4):23–24.

Good, Irving John. 1966. Speculations concerning the first ultraintelligent machine. *Advances in Computers* 6:31–88.

Hall, J. Storrs. 2008a. Engineering utopia. *Frontiers in Artificial Intelligence and Applications* 171:460.

Hall, J. Storrs. 2008b. VARIAC: an autogenous cognitive architecture. *Frontiers in Artificial Intelligence and Applications* 171:176.

Hall, John Storrs. 2007. Self-improving AI: an analysis. *Minds and Machines* 17(3):249–259.

Heylighen, Francis. 2012. Brain in a vat cannot break out. *Journal of Consciousness Studies* 19(1–2):1–2.

Hutter, Marcus. 2007. Universal algorithmic intelligence: a mathematical top-down approach. In *Artificial General Intelligence*, edited by B. Goertzer and C. Pennachin, 227–290. Berlin: Springer.

Hutter, Marcus. 2012. Can intelligence explode? *Journal of Consciousness Studies* 19(1–2):1–2.

Khalifa, Amine Ben and Roman V. Yampolskiy. 2011. GA with wisdom of artificial crowds for solving mastermind satisfiability problem. *International Journal of Intelligent Games and Simulation* 6(2):6.

Kolmogorov, A. N. 1965. Three approaches to the quantitative definition of information. *Problems of Information Transmission* 1(1):1–7.

Krauss, Lawrence M., and Glenn D. Starkman. 2004. Universal limits on computation. arXiv preprint astro-ph/0404510

Leon, J. and A. Lori. 2001. Continuous self-evaluation for the self-improvement of software. *Self-Adaptive Software* 1936:27–39.

LessWrong. 2014. Acausal Trade. Accessed. September 29. http://wiki.lesswrong.com/wiki/Acausal_trade

Levin, Leonid. 1973. Universal search problems. *Problems of Information Transmission* 9(3):265–266.

Lloyd, Seth. 2000. Ultimate physical limits to computation. *Nature* 406:1047–1054.

Loosemore, Richard, and Ben Goertzel. 2012. Why an intelligence explosion is probable. In *Singularity Hypotheses*, edited by Amnon H. Eden, James H. Moor, Johnny H. Søraker, and Eric Steinhart, 83–98. Berlin: Springer.

Mahoney, Matt. June 15, 2008. Is There a Model for RSI? *SL4*. http://www.sl4.org/archive/0806/18997.html

Mahoney, Matt. June 20, 2008. Is There a Model for RSI? *SL4*. http://www.sl4.org/archive/0806/19028.html

Mahoney, Matthew. 2010. A Model for Recursively Self Improving Programs. http://mattmahoney.net/rsi.pdf

Majot, Andrew M. and Roman V. Yampolskiy. May 23–24, 2014. AI Safety Engineering through Introduction of Self-Reference into Felicific Calculus via Artificial Pain and Pleasure. Paper presented at the 2014 IEEE International Symposium on Ethics in Science, Chicago.

Mavrogiannopoulos, Nikos, Nessim Kisserli, and Bart Preneel. 2011. A taxonomy of self-modifying code for obfuscation. *Computers and Security* 30(8):679–691.

Melkikh, Alexey V. 2014. The no free lunch theorem and hypothesis of instinctive animal behavior. *Artificial Intelligence Research* 3(4):43.

Minsky, Marvin. 1966. Artificial intelligence. *Scientific American* 215(3):247–260.

Muehlhauser, Luke, and Anna Salamon. 2012. Intelligence explosion: evidence and import. In *Singularity Hypotheses*, edited by Amnon H. Eden, James H. Moor, Johnny H. Søraker, and Eric Steinhart, 15–42. Berlin: Springer.

Nivel, Eric and Kristinn R. Thórisson. 2008. Self-Programming: Operationalizing Autonomy. Paper presented at Proceedings of the 2nd Conference on Artificial General Intelligence, Arlington, VA, March 6-9, 2009.

Nivel, Eric, Kristinn R. Thórisson, Bas R. Steunebrink, Haris Dindo, Giovanni Pezzulo, M. Rodriguez, C. Hernandez, Dimitri Ognibene, Jürgen Schmidhuber, and R. Sanz. 2013. Bounded Recursive Self-Improvement. arXiv preprint arXiv:1312.6764

Omohundro, Stephen M. September 8–9, 2007. The Nature of Self-Improving Artificial Intelligence. Paper presented at the Singularity Summit, San Francisco.

Omohundro, Steve. 2012. Rational artificial intelligence for the greater good. In *Singularity Hypotheses*, edited by Amnon H. Eden, James H. Moor, Johnny H. Søraker, and Eric Steinhart, 161–179. Berlin: Springer.

Orseau, Laurent and Mark Ring. August 3–6, 2011. Self-Modification and Mortality in Artificial Agents. Paper presented at the Fourth International Conference on Artificial General Intelligence, Mountain View, CA.

Pearce, David. 2012. The biointelligence explosion. In *Singularity Hypotheses*, edited by Amnon H. Eden, James H. Moor, Johnny H. Søraker, and Eric Steinhart, 199–238. Berlin: Springer.

Petrean, Liviu. 2010. Polymorphic and metamorphic code applications in portable executable files protection. *Acta Technica Napocensis* 51(1): 1–6.

Port, Aaron C. and Roman V. Yampolskiy. July 30–August 1, 2012. Using a GA and Wisdom of Artificial Crowds to Solve Solitaire Battleship Puzzles. Paper presented at the 17th International Conference on Computer Games (CGAMES). Louisville, KY: IEEE.

Rice, Henry Gordon. 1953. Classes of recursively enumerable sets and their decision problems. *Transactions of the American Mathematical Society* 74(2):358–366.

Sandberg, Anders. 1999. The physics of information processing superobjects: daily life among the Jupiter brains. *Journal of Evolution and Technology* 5(1):1–34.

Schaeffer, Jonathan, Neil Burch, Yngvi Bjornsson, Akihiro Kishimoto, Martin Muller, Robert Lake, Paul Lu, and Steve Sutphen. 2007. Checkers is solved. *Science* 317(5844):1518–1522.

Schaul, Tom and Juergen Schmidhuber. 2010. Metalearning. *Scholarpedia* 5(6):4650.

Schmidhuber, Juergen. 1999. A general method for incremental self-improvement and multiagent learning. In *Evolutionary Computation: Theory and Applications,* X. Yao, Ed., 81–123. Singapore: World Publishing.

Schmidhuber, Jürgen. 2004. Optimal ordered problem solver. *Machine Learning* 54(3):211–254.

Schmidhuber, Jürgen. 2005a. Completely self-referential optimal reinforcement learners. in *Artificial Neural Networks: Formal Models and Their Applications—ICANN 2005,* edited by Włodzisław Duch, Janusz Kacprzyk, Erkki Oja and Sławomir Zadrożny. *Lecture Notes in Computer Science,* 223–233. Springer Berlin Heidelberg.

Schmidhuber, Jürgen. 2005b. Gödel machines: Self-referential universal problem solvers making provably optimal self-improvements. in *Artificial General Intelligence,* edited by Ben Goertzel and Cassio Pennachin, 199–227. Berlin, Germany: Springer-Verlag.

Schmidhuber, Jürgen. 2005c. Gödel machines: Towards a technical justification of consciousness. In *Adaptive Agents and Multi-Agent Systems II,* edited by Daniel Kudenko, Dimitar Kazakov, and Eduardo Alonso, 1–23. Berlin: Springer.

Schmidhuber, Jürgen. 2007. Gödel machines: Fully self-referential optimal universal self-improvers, in *Artificial General Intelligence,* edited by Ben Goertzel and Cassio Pennachin, 199–226. Berlin, Germany: Springer-Verlag.

Schmidhuber, Jürgen. 2009. Ultimate cognition à la Gödel. *Cognitive Computation* 1(2):177–193.

Schmidhuber, Jürgen, Jieyu Zhao, and Marco Wiering. 1997. Shifting inductive bias with success-story algorithm, adaptive Levin search, and incremental self-improvement. *Machine Learning* 28(1):105–130.

Shahaf, Dafna and Eyal Amir. March 26–28, 2007. Towards a Theory of AI Completeness. Paper presented at the Eighth International Symposium on Logical Formalizations of Commonsense Reasoning (Commonsense 2007), Stanford, CA.

Shannon, C. E. 1948. A mathematical theory of communication. *Bell Systems Technical Journal* 27(3):379–423.

Smart, John M. 2009. Evo devo universe? A framework for speculations on cosmic culture. In *Cosmos and Culture: Cultural Evolution in a Cosmic Context*, edited by Mark L. Lupisella and Steven J. Dick, 201–295. Washington, DC: US Government Printing Office. NASA SP-2009-4802.

Sotala, Kaj. 2012. Advantages of artificial intelligences, uploads, and digital minds. *International Journal of Machine Consciousness* 4(01):275–291.

Sotala, Kaj and Roman V. Yampolskiy. 2015. Responses to catastrophic AGI risk: a survey. *Physica Scripta* 90:018001.

Steunebrink, B. and J. Schmidhuber. August 3–6, 2011. A Family of Gödel Machine Implementations. Paper presented at the Fourth Conference on Artificial General Intelligence (AGI-11), Mountain View, CA.

Stewart, John E. 2010. The meaning of life in a developing universe. *Foundations of Science* 15(4):395–409.

Tipler, Frank J. 1994. *The Physics of Immortality: Modern Cosmology, God, and the Resurrection of the Dead.* New York: Random House.

Turchin, Valentin F. 1986. The concept of a supercompiler. *ACM Transactions on Programming Languages and Systems (TOPLAS)* 8(3):292–325.

Turing, A. 1950. Computing machinery and intelligence. *Mind* 59(236):433–460.

Turing, Alan. 1936. On computable numbers, with an application to the Entscheidungs problem. *Proceedings of the London Mathematical Society* 2(42):230–265.

Vidal, Clément. 2013. The Beginning and the End: The Meaning of Life in a Cosmological Perspective. arXiv preprint arXiv:1301.1648

Waser, Mark R. November 9, 2014. Bootstrapping a Structured Self-Improving and Safe Autopoietic Self. Paper presented at the annual International Conference on Biologically Inspired Cognitive Architectures, Boston.

Wheeler, John Archibald. 1990. *Information, Physics, Quantum: The Search for Links.* Austin: University of Texas.

Wiedermann, Jiří. 2012a. A computability argument against superintelligence. *Cognitive Computation* 4(3):236–245.

Wiedermann, Jiri. 2012b. Is there something beyond AI? Frequently emerging, but seldom answered questions about artificial super-intelligence. In *Beyond AI: Artificial Dreams*, edited by J. Jomportl, I. Pavel, E. Zackova, M. Polak, and R. Schuster, 76–86. Pilsen, Czech: University of West Bohemia.

Wolfram, Stephen. May 14, 2002. *A New Kind of Science.* Oxfordshire, UK: Wolfram Media.

Wolpert, David H. and William G. Macready. 1997. No free lunch theorems for optimization. *IEEE Transactions on Evolutionary Computation* 1(1):67–82.

Yampolskiy, R. V. 2011a. AI-Complete CAPTCHAs as Zero Knowledge Proofs of Access to an Artificially Intelligent System. *ISRN Artificial Intelligence* no. 271878.

Yampolskiy, R. V. and E. L. B. Ahmed. Wisdom of artificial crowds algorithm for solving NP-hard problems. *International Journal of Bio-Inspired Computation (IJBIC)* 3(6):358–369.

Yampolskiy, Roman, Peter Anderson, Jonathan Arney, Vladimir Misic, and Trevor Clarke. September 24, 2004. Printer Model Integrating Genetic Algorithm for Improvement of Halftone Patterns. Paper presented at the Western New York Image Processing Workshop (WNYIPW), IEEE Signal Processing Society, Rochester, NY.

Yampolskiy, Roman, Gyuchoon Cho, Richard Rosenthal, and Marina Gavrilova. 2012. Experiments in Artimetrics: Avatar Face Recognition. *Transactions on Computational Science* 16:77–94.

Yampolskiy, Roman and Joshua Fox. 2012. Safety engineering for artificial general intelligence. *Topoi*. October 2013, 32(2): 217–226.

Yampolskiy, Roman and Marina Gavrilova. 2012. Artimetrics: biometrics for artificial entities. *IEEE Robotics and Automation Magazine (RAM)* 19(4):48–58.

Yampolskiy, Roman V. 2011b. Construction of an NP Problem with an Exponential Lower Bound. Arxiv preprint arXiv:1111.0305

Yampolskiy, Roman V. April 21–22, 2012a. AI-Complete, AI-Hard, or AI-Easy—Classification of Problems in AI. Paper presented at the 23rd Midwest Artificial Intelligence and Cognitive Science Conference, Cincinnati, OH.

Yampolskiy, Roman V. April 21–22, 2012b. Computing Partial Solutions to Difficult AI Problems. Paper presented at the 23rd Midwest Artificial Intelligence and Cognitive Science Conference, Cincinnati, OH.

Yampolskiy, Roman V. 2012c. Leakproofing singularity-artificial intelligence confinement problem. *Journal of Consciousness Studies (JCS)* 19(1–2):194–214.

Yampolskiy, Roman V. 2013a. Artificial intelligence safety engineering: why machine ethics is a wrong approach. In *Philosophy and Theory of Artificial Intelligence*, edited by Vincent C. Müller, 389–396. Berlin: Springer.

Yampolskiy, Roman V. 2013b. Efficiency theory: a unifying theory for information, computation and intelligence. *Journal of Discrete Mathematical Sciences and Cryptography* 16(4–5):259–277.

Yampolskiy, Roman V. 2013c. What to do with the singularity paradox? In *Philosophy and Theory of Artificial Intelligence*, 397–413. Berlin: Springer.

Yampolskiy, Roman V. 2013d. Turing test as a defining feature of AI-Completeness. In *Artificial Intelligence, Evolutionary Computing and Metaheuristics*, edited by Xin-She Yang, 3–17. Berlin: Springer.

Yampolskiy, Roman V. 2014a. The Universe of Minds. arXiv preprint arXiv:1410.0369

Yampolskiy, Roman V. 2014b. Utility function security in artificially intelligent agents. *Journal of Experimental and Theoretical Artificial Intelligence (JETAI)* 26(3): 373–389.

Yampolskiy, Roman V., Leif Ashby, and Lucas Hassan. 2012. Wisdom of artificial crowds—a metaheuristic algorithm for optimization. *Journal of Intelligent Learning Systems and Applications* 4(2):98–107.

Yampolskiy, Roman V. and Joshua Fox. 2012. Artificial general intelligence and the human mental model. In *Singularity Hypotheses: A Scientific and Philosophical Assessment*, edited by Amnon H. Eden, James H. Moor, Johnny H. Søraker, and Eric Steinhart, 129. Berlin: Springer.

Yonck, Richard. 2012. Toward a standard metric of machine intelligence. *World Future Review* 4(2):61–70.

Yudkowsky, Eliezer. 2007. Levels of organization in general intelligence. In *Artificial General Intelligence*, edited by B. Goertzer and C. Pennachin, 389–501. Berlin: Springer.

Yudkowsky, Eliezer. December 1, 2008. Recursive Self-Improvement. Less Wrong. http://lesswrong.com/lw/we/recursive_selfimprovement/ Accessed September 29, 2014.

Yudkowsky, Eliezer. 2010. *Timeless Decision Theory*. San Francisco: Singularity Institute.

Yudkowsky, Eliezer. 2013. *Intelligence Explosion Microeconomics*. MIRI Technical Report 2013-1. Berkeley, CA: Machine Intelligence Research Institute. http://www.intelligence.org/files/IEM.pdf

Yudkowsky, Eliezer. 2014. *The Procrastination Paradox* (Brief technical note). MIRI Technical Report. https://intelligence.org/files/ProcrastinationParadox.pdf

Yudkowsky, Eliezer and Robin Hanson. 2008. *The Hanson-Yudkowsky AI-Foom Debate*. MIRI Technical Report. http://intelligence.org/files/AIFoomDebate.pdf

Yudkowsky, Eliezer and Marcello Herreshoff. 2013. *Tiling Agents for Self-Modifying AI, and the Löbian Obstacle*. MIRI Technical Report. http://intelligence.org/files/TilingAgentsDraft.pdf

Yudkowsky, Eliezer S. 2001. General Intelligence and Seed AI—Creating Complete Minds Capable of Open-Ended Self-Improvement. http://singinst.org/ourresearch/publications/GISAI/

Yudkowsky, Eliezer S. May 2004. Coherent Extrapolated Volition. Singularity Institute for Artificial Intelligence. http://singinst.org/upload/CEV.html

CHAPTER 6

Singularity Paradox and What to Do About It*

6.1 INTRODUCTION TO THE SINGULARITY PARADOX

Many philosophers, futurologists, and artificial intelligence (AI) researchers (Solomonoff 1985; Bostrom 2006; Yudkowsky 2007, 2008; Hawking 1998; Kurzweil 2005; "Tech Luminaries" 2008) have conjectured that in the next 20 to 200 years a machine capable of at least human-level performance on all tasks will be developed. Because such a machine would, among other things, be capable of designing the next generation of even smarter intelligent machines, it is generally assumed that an intelligence explosion will take place shortly after such a technological self-improvement cycle begins (Good 1966). Although specific predictions regarding the consequences of such an intelligence singularity are varied from potential economic hardship (Hanson 2008) to the complete extinction of humankind (Yudkowsky 2008; Bostrom 2006), many of the involved researchers agree that the issue is of utmost importance and needs to be seriously addressed (Chalmers 2010).

Investigators concerned with the existential risks posed to humankind by the appearance of superintelligence often describe what I shall call a *singularity paradox* (SP) as their main reason for thinking that humanity might

* Reprinted from Roman V. Yampolskiy, *Studies in Applied Philosophy, Epistemology and Rational Ethics* 5:397–413, 2013, with kind permission of Springer Science and Business Media. Copyright 2013, Springer Science and Business Media.

be in danger. Briefly, SP could be described as follows: Superintelligent machines are feared to be too dumb to possess common sense.

SP is easy to understand via some commonly cited examples. Suppose that scientists succeed in creating a superintelligent machine and order it to "make all people happy." Complete happiness for humankind is certainly a noble and worthwhile goal, but perhaps we are not considering some unintended consequences of giving such an order. Any human immediately understands what is meant by this request; a nonexhaustive list may include making all people healthy, wealthy, beautiful, and talented and giving them loving relationships and novel entertainment. However, many alternative ways of making all people happy could be derived by a superintelligent machine. For example:

- Killing all people trivially satisfies this request as with 0 people, all of them are happy.
- Forced lobotomies for every man, woman, and child might also accomplish the same goal.
- A simple observation that happy people tend to smile may lead to forced plastic surgeries to affix permanent smiles to all human faces.
- A daily cocktail of cocaine, methamphetamine, methylphenidate, nicotine, and 3,4-methylenedioxymethamphetamine, better known as ecstasy, may do the trick.

An infinite number of other approaches to accomplish universal human happiness could be derived. For a superintelligence, the question is simply which one is fastest/cheapest (in terms of computational resources) to implement. Such a machine clearly lacks common sense, hence the paradox.

We want our machines to do what we want, not what we tell them to do, but as bugs in our programs constantly teach us, this is not a trivial task. The next section of this chapter presents an overview of different approaches proposed for either dealing with the SP or avoiding it all together. In particular, many of the reviewed ideas address a generalized version of the SP that could be stated as follows: We build this machine to have a property *X*, but it actually does ~*X*. Here, *X* could stand for the original goal of happiness or it could represent any of its components, such as security, prosperity, socialization, and so on.

6.2 METHODS PROPOSED FOR DEALING WITH SP

6.2.1 Prevention from Development

6.2.1.1 Fight Scientists

One of the earliest and most radical critics of the upcoming singularity was Theodore Kaczynski, a Harvard-educated mathematician also known as the Unabomber. His solution to prevent singularity from ever happening was a bloody multiyear terror campaign against university research labs across the United States. In his 1995 manifesto, Kaczynski explains his negative views regarding the future of humankind dominated by machines: "If the machines are permitted to make all their own decisions, we can't make any conjectures as to the results, because it is impossible to guess how such machines might behave. We only point out that the fate of the human race would be at the mercy of the machines" (Kaczynski 1995, 79).

An even more violent outcome is prophesized, but not advocated, by Hugo de Garis (2005), who predicts that the issue of building superintelligent machines will split humanity into two camps, eventually resulting in a civil war over the future of singularity research: "I believe that the ideological disagreements between these two groups on this issue will be so strong, that a major … war, killing billions of people, will be almost inevitable before the end of the 21st century" (2005, 234).

6.2.1.2 Restrict Hardware and Outlaw Research

Realizing the potential dangers of superintelligent computers, Anthony Berglas proposed a legal solution to the problem. He suggested outlawing production of more powerful processors, essentially stopping Moore's law in its tracks and consequently denying necessary computational resources to self-improving artificially intelligent machines (Berglas 2009). Similar laws aimed at promoting human safety have been passed banning research on cloning of human beings and development of biological (1972 Biological Weapons Convention), chemical (1993 Chemical Weapons Convention), and nuclear weaponry. Berglas's idea may be interesting in terms of its shock value, which in turn may attract more attention to the dangers of the SP. Here is what Berglas suggested in his own words: "a radical solution, namely to limit the production of ever more powerful computers and so try to starve any AI of processing power. This is urgent, as computers are already almost powerful enough to host an artificial intelligence. … One major problem is that we may already have sufficient power in general purpose computers to support intelligence" (Berglas 2009).

Alternatively, restrictions could be placed on the intelligence an AI may possess to prevent it from becoming superintelligent (Gibson 1984) or legally require that its memory be erased after every job (Benford 1988). Similarly, Bill Joy advocates for relinquishment of superintelligence research and even suggests how enforcement of such a convention could be implemented (Joy 2000): "Enforcing relinquishment will require a verification regime similar to that for biological weapons, but on an unprecedented scale. ... Verifying compliance will also require that scientists and engineers adopt a strong code of ethical conduct, resembling the Hippocratic oath, and that they have the courage to whistleblow as necessary, even at high personal cost."

For enforcement of such technology, restricting laws will not be trivial unless the society as a whole adopts an Amish-like, technology-free, lifestyle.

6.2.1.3 Singularity Steward

Ben Goertzel, a computer scientist, has proposed creation of a "big brother AI" monitoring system he calls the "singularity steward." The goal of the proposed system is to monitor the whole world with the specific aim of preventing development of any technology capable of posing a risk to humanity, including superintelligent machines (Goertzel 2004b). Goertzel believes that creation of such a system is feasible and would safeguard humanity against preventable existential risks. Goertzel (2004b) also claims that "in the AI Big Brother case, one doesn't want the AI to be self-modifying and self-improving—one wants it to remain stable. ... One needs to make it a bit smarter than humans, but not too much—and one needs to give it a goal system focused on letting itself and humans remain as much the same as possible."

6.2.2 Restricted Deployment

6.2.2.1 AI-Box

A common theme in singularity discussion forums is the possibility of simply keeping a superintelligent agent in sealed hardware to prevent it from doing any harm to humankind. Such ideas originate with scientific visionaries such as Eric Drexler, who has suggested confining transhuman machines so that their outputs could be studied and used safely (Drexler 1986). The general consensus on such an approach among researchers seems to be that such confinement is impossible to successfully maintain. For example, Vernor Vinge has strongly argued against the case of physical confinement (Vinge 1993): “Imagine yourself locked in your home with only limited data access to the outside, to your masters. If those masters thought at a rate—say—one million times slower than you, there is little doubt that over a period of years (your time) you could come up with ‘helpful advice’ that would incidentally set you free.”

Likewise, David Chalmers, a philosopher, has stated that confinement is impossible because any useful information we would be able to extract from the AI will affect us, defeating the purpose of confinement (Chalmers 2010). However, the researcher who did the most to discredit the idea of the so-called AI-Box is Eliezer Yudkowsky, who has actually performed AI-Box “experiments” in which he demonstrated that even human-level intelligence is sufficient to escape from an AI-Box (Yudkowsky 2002). In a series of five experiments, Yudkowsky challenged different individuals to play a role of a gatekeeper to a superintelligent agent (played by Yudkowsky himself) trapped inside an AI-Box and was successful in securing his release in three of five trials via nothing more than a chat interface (Yudkowsky 2002).

6.2.2.2 *Leakproof Singularity*

In 2010, David Chalmers proposed the idea of a “leakproof” singularity. He suggests that, for safety reasons, first AI systems be restricted to simulated virtual worlds until their behavioral tendencies can be fully understood under the controlled conditions. Chalmers argues that even if such an approach is not foolproof, it is certainly safer than building AI in physically embodied form. However, he also correctly observes that a truly leakproof system in which no information is allowed to leak out from the simulated world into our environment “is impossible, or at least pointless” (Chalmers 2010, 38) because we cannot interact with the system or even observe it. Chalmers’s discussion of the leakproof singularity is an excellent introduction to the state-of-the-art thinking in the field: “The obvious suggestion is that we should first create AI and AI+ systems in virtual worlds: simulated environments that are themselves realized inside a computer. Then an AI will have free reign within its own world without being able to act directly on ours” (Chalmers 2010, 37).

6.2.2.3 *Oracle AI*

Nick Bostrom, a futurologist, has proposed (Bostrom 2008) an idea for an oracle AI (OAI), which would only be capable of answering questions. It is easy to elaborate and see that a range of different OAIs is possible. From advanced OAIs capable of answering any question to domain expert AIs capable of answering Yes/

No/Unknown to questions on a specific topic. It is claimed that an OAI could be used to help humankind build a safe, unrestricted superintelligent machine.

6.2.2.4 AI Confinement Protocol

I proposed a formalized notion of AI confinement protocol (cp) that represents "AI-Boxing" as a computer security challenge. I define the AI confinement problem (AICP) as the challenge of restricting an artificially intelligent entity to a confined environment from which it cannot exchange information with the outside environment via legitimate or covert channels if such information exchange was not authorized by the confinement authority. An AI system that succeeds in violating the CP protocol is said to have escaped (Yampolskiy 2012).

The proposed protocol is based on the idea of asking what are called "safe questions" with restricted answers. A *safe question* is defined as one for which a human being could find an answer without any help from superintelligence, all on his or her own merit. Why is this useful? Suppose that a scientist working on a cure for cancer is looking at two potential cures. Each one could be developed and tested without assistance from the AI but would take about 3 years of hard work each. Which one is the scientist to try first if he or she believes that they have about an equal chance of working? If a wrong option is chosen to be tested first, humanity will still obtain a cure for cancer, but it will happen 3 years later. What if we could ask the AI to suggest which option to try first?

This question is safe for a number of reasons. First, each potential answer has an equal probability of being correct. Second, a human being could answer it without assistance from the AI, so asking the AI for help will only speed progress but in no way change the outcome. It is essentially equal to being lucky at guessing multiple-choice answers. Finally, the answer to this question could be encoded in a single bit, making it hard to hide additional messages in the provided answer.

To make sure that a potential question is safe, a group of experts could be used to review it. The experts should all be trained AI safety engineers, meaning that they are familiar with the design of the AI and its confinement environment as well as the latest developments in machine ethics (Allen, Wallach, and Smit 2006; Anderson and Anderson 2007; Hall 2000; Moor 2006; Tonkens 2009). Experts may also need to be trained in computer psychology, a currently nonexistent profession that might become a reality in the future (Epstein 1997). An existing discipline that might be of greatest help for training of AI question review experts is artimetrics, a field of study I proposed that identifies, classifies, and authenticates AI agents, robots, and virtual reality avatars for security purposes (Yampolskiy 2007; Yampolskiy and Govindaraju 2007, 2008; Gavrilova and Yampolskiy 2010).

6.2.3 Incorporation into Society

6.2.3.1 Law and Economics

Robin Hanson has suggested that as long as future intelligent machines are law abiding, they should be able to coexist with humans (Hanson 2009): "In the early to intermediate era when robots are not vastly more capable than humans, you'd want peaceful law-abiding robots as capable as possible, so as to make productive partners. You might prefer they dislike your congestible goods, like your scale-economy goods, and vote like most voters, if they can vote."

Similarly, Hans Moravec puts his hopes for humanity in the hands of the law. He sees forcing cooperation from the robot industries as the most important security guarantee for humankind and integrates legal and economic measures into his solution (Joy 2000): "In a completely free marketplace, superior robots would surely affect humans. … Robotic industries would compete vigorously among themselves for matter, energy, and space, incidentally driving their price beyond human reach. … Judiciously applied, governmental coercion could support human populations in high style on the fruits of robot labor, perhaps for a long while."

Robin Hanson, an economist, agrees: "Robots well-integrated into our economy would be unlikely to exterminate us" (Hanson 2008, 50). Similarly, Steve Omohundro uses microeconomic theory to speculate about the driving forces in the behavior of superintelligent machines. He argues that intelligent machines will want to self-improve, be rational, preserve their utility functions, prevent counterfeit utility, acquire resources and use them efficiently, and protect themselves. He believes that machines' actions will be governed by rational economic behavior (Omohundro 2007, 2008).

Mark Waser suggests an additional "drive" to be included in the list of behaviors predicted to be exhibited by the machines (Waser 2010b). Namely, he suggests that evolved desires for cooperation and being social are part of human ethics and are a great way of accomplishing goals, an idea also analyzed by Joshua Fox and Carl Shulman (2010). Bill Hibbard adds the desire for maintaining the social contract toward equality as a component of ethics for superintelligent machines (Hibbard 2005a), and J. Storrs Hall argues for incorporation of moral codes into the design (Hall 2000). In general, ethics for superintelligent machines is one of the most fruitful areas of research in the field of singularity research, with numerous publications appearing every year (Shulman, Jonsson, and Tarleton 2009; Bostrom and Yudkowsky 2011; Bostrom 2006; Sotala 2009; Shulman, Tarleton, and Jonsson 2009; Waser 2010a; Bugaj and Goertzel 2007).

6.2.3.2 Religion for Robots

Robert Geraci, a theologian, has researched similarities between different aspects of technological singularity and the world's religions (Geraci 2006). In particular, in his work on apocalyptic AI (Geraci 2008), he observes the many commonalities in the works of biblical prophets like Isaiah and the prophets of the upcoming technological singularity, such as Ray Kurzweil or Hans Moravec. All promise freedom from disease, immortality, and purely spiritual (software) existence in the kingdom come (Virtual Reality). More interestingly, Geraci (2007) argues that to be accepted into the society as equals, robots must convince most people that they are conscious beings. Geraci believes that an important component for such attribution is voluntary religious belief. Just like some people choose to believe in a certain religion, so will some robots. In fact, one may argue that religious values may serve the goal of limiting the behavior of superintelligences to those acceptable to society just like they do for many people. Here is how Geraci motivates his argument (Geraci 2007): "If robots become conscious, they may desire entrance into our society. … If no robots can enter into our religious lives, then I suspect we will deny them all equal and near-equal status in our culture. … To qualify as 'persons,' … some of them need to be religious—and by choice, not deliberate programming."

Adherents of Eastern religions are even more robot friendly and in general assume that robots will be happy to serve society and pose no danger. For example, Japan's Fumio Hara thinks that if "you are a good, kind person to the robot, the robot will become kind in return" (Menzel and D'Aluisio 2001, 76). Another eminent Japanese scientist, Shigeo Hirose, believes that robots "can be saints-intelligent and unselfish" (Menzel and D'Aluisio 2001, 89). Overall, convincing robots to worship humans as gods may be a valid alternative to friendly and humane AI systems.

6.2.3.3 Education

David Brin, in a work of fiction, has proposed that smart machines should be given humanoid bodies and from inception raised as our children and taught the same way we were (Brin 1987). Instead of programming machines explicitly to follow a certain set of rules, they should be given the capacity to learn and should be immersed in human society with its ethical and cultural rules.

6.2.4 Self-Monitoring

6.2.4.1 Hard-Coded Rules

Probably the earliest and the best-known solution for the problem of intelligent machines was proposed by Isaac Asimov, a biochemist and a science fiction writer, in the early 1940s. The so-called Three Laws of Robotics are almost universally known and have inspired numerous imitations as well as heavy critique (Gordon-Spears 2003; McCauley 2007; Weld and Etzioni 1994; Pynadath and Tambe 2002). The original laws as given by Asimov are as follows (Asimov 1942):

1. A robot may not injure a human being or, through inaction, allow a human being to come to harm.
2. A robot must obey orders given to it by human beings except where such orders would conflict with the First Law.
3. A robot must protect its own existence as long as such protection does not conflict with either the First or Second Law.

Asimov later expanded the list to include a number of additional laws ("Three Laws" 2015):

1. *Zeroth Law:* A robot may not harm humanity or through inaction allow humanity to come to harm.
2. *Minus-One Law:* A robot may not harm sentience or through inaction allow sentience to come to harm.
3. *Fourth Law:* A robot must establish its identity as a robot in all cases.
4. Alternate *Fourth Law:* A robot must reproduce unless such reproduction would interfere with the First or Second or Third Law.
5. *Fifth Law:* A robot must know it is a robot.

Continuing Asimov's work, rule-based standards of behavior for robots have been recently proposed by South Korea's Ministry of Commerce,

Industry, and Energy. In 2007, a Robot Ethics Charter, which sets ethical guidelines concerning robot functions, has been adopted. In Europe, EURON (European Robotics Research Network) also announced plans to develop guidelines for robots in five areas: safety, security, privacy, traceability, and identifiability. Japan's Ministry of Economy, Trade, and Industry has issued policies regarding robots in homes and how they should behave and be treated (Sawyer 2007).

6.2.4.2 Chaining God

Stuart Armstrong proposed that trustworthiness of a superintelligent system could be monitored via a chain of progressively less-powerful AI systems all the way down to the human level of intelligence. The proposed "chain" would allow people to indirectly monitor and perhaps control the ultraintelligent machine. However, Armstrong himself acknowledges a number of limitations for the proposed method: The meaning of communication could be lost from one AI level to the next, or AI links in the chain may not be able to reliably judge the trustworthiness of a more intelligent entity. In such cases, the proposed solution is to shut down all AI and to start building the chain from scratch. Here is how Armstrong describes the chain: "If there were an entity, … just below the level of the GodAI …, [it] might be powerful enough, and smart enough, to conclude the GodAI was trustworthy. Then, assuming a level of AI intelligence just below the [entity] that could check up on them, the message of trust could be passed down, eventually reaching us" (Armstrong 2007, 3).

6.2.4.3 Friendly AI

To protect humankind against unintended consequences of superintelligent machines, Eliezer Yudkowsky, an AI researcher, has suggested that any AI system under development should be "friendly" to humanity (Yudkowsky 2008). Friendliness, according to Yudkowsky, could be defined as looking out for the best interests of humankind. To figure out what humankind is really interested in, the design of friendly AI (FAI) should be done by specialized AIs. Such seed AI (Yudkowsky 2001b) systems will first study human nature and then produce a friendly superintelligence humanity would want if it was given sufficient time and intelligence to arrive at a satisfactory design, our coherent extrapolated volition (CEV) (Yudkowsky 2004). Yudkowsky is not the only researcher working on the problem of extracting and understanding human desires. Tim Freeman has also attempted to formalize a system capable of such "wish mining" but in the context of "compassionate" and "respectful" plan development by AI systems (Freeman 2009).

For friendly self-improving AI systems, a desire to pass friendliness as a main value to the next generation of intelligent machines should be a fundamental drive. Yudkowsky also emphasizes the importance of the "first mover advantage": The first superintelligent AI system will be powerful enough to prevent any other AI systems from emerging, which might protect humanity from harmful AIs. Here is how Yudkowsky himself explains FAI and CEV:

> The term "Friendly AI" refers to the production of human-benefiting, non-human-harming actions in Artificial Intelligence systems that have advanced to the point of making real-world plans in pursuit of goals. (Yudkowsky 2001a, 2)
>
> … Our coherent extrapolated volition is our wish if we knew more, thought faster, were more the people we wished we were, had grown up farther together; where the extrapolation converges rather than diverges, where our wishes cohere rather than interfere; extrapolated as we wish that extrapolated, interpreted as we wish that interpreted. (Yudkowsky 2004, 6)

6.2.4.4 Humane AI

Ben Goertzel, a frequent critic of FAI (Goertzel 2006), has proposed a variation on the theme he calls a humane AI. He believes it is more feasible to install AI with general properties such as compassion, choice, and growth than with specific properties like friendliness to humans (Goertzel 2006). In Goertzel's own words (Goertzel 2004b): "In Humane AI, one posits as a goal, … the development of AI's that display the qualities of 'humaneness,' … as a kind of ethical principle, where the principle is: 'Accept an ethical system to the extent that is agrees with the body of patterns known as 'humaneness.'"

6.2.4.5 Emotions

Bill Hibbard believes that the design of superintelligent machines needs to incorporate emotions that can guide the process of learning and self-improvement in such machines. In his opinion, machines should love us as their most fundamental emotion; consequently, they will attempt to make us happy and prosperous. He states: "So in place of laws constraining the behavior of intelligent machines, we need to give them emotions that can guide their learning of behaviors. They should want us to be happy and prosper, which is the emotion we call love" (Hibbard 2001, 12).

Others have also argued for the importance of emotions, for example, Mark Waser wrote: "Thinking machines need to have analogues to emotions like fear and outrage that create global biases towards certain actions and reflexes under appropriate circumstances" (Waser 2010b, 174).

6.2.5 Indirect Solutions

6.2.5.1 Why They May Need Us

Continuing with the economic model of supply and demand, it is possible to argue that the superintelligent machines will need humans and therefore not exterminate humanity (but still might treat it less than desirably). For example, in the movie *Matrix,* machines need the heat from our bodies as energy. It is not obvious from the movie why this would be an efficient source of energy, but we can certainly think of other examples.

Friendly AI is attempting to replicate what people would refer to as "common sense" in the domain of plan formation (Yudkowsky 2005). Because only humans know what it is like to be a human (Nagel 1974), the friendly machines would need people to provide that knowledge, to essentially answer the question: "What would a human do (WWHD)?"

Alan Turing, in "Intelligent Machinery, a Heretical Theory," argued that humans can do something machines cannot, namely, overcome limitations of Godel's incompleteness theorem (Turing 1996). Here is what Turing said on this matter: "By Godel's famous theorem, or some similar argument, one can show that however the machine is constructed there are bound to be cases where the machine fails to give an answer, but a mathematician would be able to" (Turing 1996, 256).

Another area of potential need for assistance from human beings for machines may be deduced from some peer-reviewed experiments showing that human consciousness can affect random number generators and other physical processes (Bancel and Nelson 2008). Perhaps ultraintelligent machines will want that type of control or some more advanced technology derivable from it.

As early as 1863, Samuel Butler argued that the machines will need us to help them reproduce:

> They cannot kill us and eat us as we do sheep; they will not only require our services in the parturition of their young (which branch of their economy will remain always in our hands), but also in feeding them, in setting them right when they are sick, and burying their dead or working up their corpses into new machines. … The fact is that our interests are inseparable from theirs, and theirs from ours. Each race is dependent upon the other for innumerable benefits, and, until the reproductive organs of the machines have been developed in a manner which we are hardly yet able to conceive, they are entirely dependent upon man for even the continuance of their species. It is true that these organs may be ultimately developed, inasmuch as man's interest lies in that direction; there is nothing which our infatuated race would desire more than to see a fertile union between two steam engines; it is true that machinery is even at this present time employed in begetting machinery, in becoming the parent of machines often after its own kind, but the days of flirtation, courtship, and matrimony appear to be very remote, and indeed can hardly be realized by our feeble and imperfect imagination. (Butler 1863, 184)

A set of anthropomorphic arguments is also often made. They usually go something like the following: By analyzing human behavior, we can see some reasons for a particular type of intelligent agent not to exterminate a less-intelligent life form. For example, humankind does not need elephants, and we are smarter and certainly capable of wiping them out, but instead we spend lots of money and energy preserving them. Why? Is there something inherently valuable in all life-forms? Perhaps their DNA is a great source of knowledge that we may later use to develop novel medical treatments? Or, maybe their minds could teach us something? Maybe the fundamental rule implanted in all intelligent agents should be that information should never be destroyed. As each living being is certainly packed with unique information, this would serve as a great guiding principle in all decision making. Similar arguments could be made about the need of superintelligent machines to have cute human pets, a desire for companionship with other intelligent species, or a milliard other human needs. For example, Mark Waser, a proponent of teaching the machines universal ethics (Waser 2008), which only exist in the context of society, suggested that we should "convince our super-intelligent AIs that it is in their own self-interest to join ours."

6.2.5.2 Let Them Kill Us

Some scientists are willing to give up on humanity all together in the name of a greater good they claim ultraintelligent machines will bring (Dietrich 2007). They see machines as the natural next step in evolution and believe that humanity has no right to stand in the way of progress. Essentially, their position is to let the machines do what they want, they are the future, and lack of humanity is not necessarily a bad thing. They may see the desire to keep humanity alive as nothing but a self-centered bias of *Homo sapiens*. Some may even give reasons why humanity is undesirable to nature, such as the environmental impact on Earth and later maybe the cosmos at large. According to some of the proponents of the "let-them-kill-us" philosophy: "Humans should not stand in the way of a higher form of evolution. These machines are godlike. It is human destiny to create them," believes Hugo de Garis (1999).

6.2.5.3 War Against the Machines

Amazingly, as early as 1863, Samuel Butler wrote about the need for a violent struggle against machine oppression:

> Day by day, however, the machines are gaining ground upon us; day by day we are becoming more subservient to them; … the time will come when the machines will hold the real supremacy over the world and its inhabitants is what no person of a truly philosophic mind can for a moment question. Our opinion is that war to the death should be instantly proclaimed against them. Every machine of every sort should be destroyed by the well-wisher of his species. Let there be no exceptions made, no quarter shown; let us at once go back to the primeval condition of the race. If it be urged that this is impossible under the present condition of human affairs, this at once proves that the mischief is already done, that our servitude has commenced in good earnest, that we have raised a race of beings whom it is beyond our power to destroy, and that we are not only enslaved but are absolutely acquiescent in our bondage. (Butler 1863, 185)

6.2.5.4 *If You Cannot Beat Them, Join Them*

An alternative vision for the postsingularity future of humanity could be summarized as: "If you cannot beat them, join them." A number of prominent scientists have suggested pathways for humanity to be able to keep up with superintelligent machines by becoming partially or completely merged with our engineered progeny. Ray Kurzweil is an advocate of a process known as uploading, in which the mind of a person is scanned and copied into a computer (Kurzweil 2005). The specific pathway to such scanning is not important, but suggested approaches include advanced brain-computer interfaces (BCIs), brain scanning, and nanobots. A copied human could either reside in a robotic body or in virtual reality. In any case, superior computational resources in terms of processing speed and memory become available to such an uploaded human, making it feasible for the person to keep up with superintelligent machines.

A slightly less-extreme approach is proposed by Kevin Warwick, who also agrees that we will merge with our machines but via direct integration of our bodies with them. Devices such as brain implants will give "cyborgs" computational resources necessary to compete with the best of the machines. Novel sensors will provide sensual experiences beyond the five we are used to operating with. A human being with direct uplink to the wireless Internet will be able to instantaneously download necessary information or communicate with other cyborgs (Warwick 2003). Both Kurzweil and Warwick attempt to analyze potential consequences of humanity joining the machines and come up with numerous fascinating predictions. The one aspect they agree on is that humanity will never be the same. Peter Turney suggests an interesting twist on the "fusion" scenario: "One approach to controlling a SIM would be to link it directly to a human brain. If the link is strong enough, there is no issue of control. The brain and the computer are one entity; therefore, it makes no sense to ask who is controlling whom" (Turney 1991, 3).

6.2.5.5 *Other Approaches*

I have reviewed some of the most prominent and frequently suggested approaches for dealing with the SP, but many other approaches and philosophical viewpoints are theoretically possible (Sotala and Yampolskiy 2015). Many of them would fall into the singularity "denialist" camp, accepting the following statement by Jeff Hawkins ("Tech Luminaries" 2008): "There will be no singularity or point in time where the technology itself runs away from us." He further elaborates: "Exponential growth requires the exponential consumption of resources (matter, energy, and time), and there are always limits to this. Why should we think intelligent machines would be different? We will build machines that are more 'intelligent' than humans and this might happen quickly, but there will be no singularity, no runaway growth in intelligence." A recent report from the Association for the Advancement of Artificial Intelligence (AAAI) presidential panel on long-term AI futures outlines similar beliefs held by the majority of the participating AI scientists: "There was overall skepticism about the prospect of an intelligence explosion as well as of a 'coming singularity,' and also about the large-scale loss of control of intelligent systems" (Horvitz and Selman 2009).

Others may believe that we might get lucky and even if we do nothing, the superintelligence will turn out to be friendly to us and possess some human characteristics. Perhaps this will happen as a side effect of being (directly or indirectly) designed by human engineers, who will, maybe subconsciously, incorporate such values into their designs or, as Douglas Hofstadter put it ("Tech Luminaries" 2008): "Perhaps these machines—our 'children'—will be vaguely like us and will have culture similar to ours." Yet others think that superintelligent machines will be neutral toward us. John Casti thinks that ("Tech Luminaries" 2008) "machines will become increasingly uninterested in human affairs just as we are uninterested in the affairs of ants or bees. But it's more likely than not in my view that the two species will comfortably and more or less peacefully coexist." Both Peter Turney (1991) and Alan Turing (1950) suggested that giving machines an ability to feel pleasure and pain will allow us to control them to a certain degree and will assist in machine learning. Unfortunately, teaching machines to feel pain is not an easy problem to solve (Bishop 2009; Dennett 1978).

Finally, one can simply deny that the problem exists by questioning either the possibility of the technological singularity or not accepting that it leads to the SP. Perhaps one can believe that a superintelligent machine by its very definition will have at least as much common sense as an average human and will consequently act accordingly.

6.3 ANALYSIS OF SOLUTIONS

Table 6.1 provides a summary of the methods described in this chapter proposed to either directly or indirectly address the problem we have named the SP. I have categorized the proposed solutions into five broad categories: prevention of development, restricted deployment,

TABLE 6.1 Summary of the Potential Solution Methods

Category	Methodology	Investigated by	Year
Prevention of development	Fight scientists	Ted Kaczynski	1995
	Outlaw research	Bill Joy	2000
	Restrict hardware	Anthony Berglas	2009
	Singularity steward	Ben Goertzel	2004
Restricted deployment	AI-Boxing	Eric Drexler, Eliezer Yudkowsky	2002
	Leakproofing	David Chalmers	2010
	Oracle AI	Nick Bostrom	2008
	AI-Confinement	Roman V. Yampolskiy	2011
Incorporation into society	Economic	Robin Hanson	2008
	Legal	H. Moravec, R. Hanson, S. Omohundro	2007
	Religious	Robert Geraci	2007
	Ethical/social	Mark Waser, Joshua Fox, Carl Shulman	2008
	Moral	J. Storrs Hall	2000
	Equality	Bill Hibbard	2005
	Education	David Brin	1987
Self-monitoring	Rules to follow	Isaac Asimov	1942
	Friendly AI	Eliezer Yudkowsky	2001
	Emotions	Bill Hibbard	2001
	Chaining	Stuart Armstrong	2007
	Humane AI	Ben Goertzel	2004
	Compassionate AI	Tim Freeman	2009
Other solutions	They will need us	Alan Turing	1950
	War against machines	Samuel Butler	1863
	Join them	Ray Kurzweil, Kevin Warwick	2003
	Denialism	Jeff Hawkins	2008
	Do nothing	Douglas Hofstadter, John Casti	2008
	Pleasure and pain	Peter Turney	1991
	Let them kill us	Hugo de Garis, Eric Dietrich	2005
	Fusion of humans and AI	Peter Turney	1991
	Reproductive control	Samuel Butler	1863

incorporation into society, self-monitoring, and indirect solutions. Such grouping makes it easier both to understand the proposed methods and to analyze them as a set of complete measures. I review each category and analyze it in terms of feasibility of accomplishing the proposed actions and, more important, for evaluating the likelihood of the method succeeding if implemented.

The violent struggle against scientific establishment, outlawing AI research, and placing restrictions on development and sale of hardware components are all part of an effort to prevent superintelligent machines from ever coming into existence and to some extent are associated with the modern Luddite movement. Given the current political climate, complex legal system, and economic needs of the world's most developed countries, it is highly unlikely that laws will be passed to ban computer scientists either from researching AI systems or from developing and selling faster processors. Because for this methodology to work the ban needs to be both global and enforceable, it will not work as there is no global government to enforce such a law or to pass it in the first place. Even if such a law were passed, there is always a possibility that some rogue scientist somewhere will simply violate the restrictions, making them at best a short-term solution.

An idea for an automated monitoring system (also known as "big brother AI") is as likely to be accepted by humanity as the legal solution analyzed previously. It also presents the additional challenge of technological implementation, which as far as I can tell would be as hard to make "humanity safe" as a full-blown singularity-level AI system. Provided that the system would have to be given legal rights to control people, Martha Moody said: "Sometimes the cure is worse than the disease." Finally, as for the idea of violent struggle, it may come to be, as suggested by Hugo de Garis (2005), but I will certainly not advocate such an approach or will even consider it as a real solution.

Restricting access of superintelligent machines to the real world is a commonly proposed solution to the SP problem. AI-Boxes, leakproofing, and restricted question-answering-only systems (known as oracle AIs) are just some of the proposed methods for accomplishing that. Although much skepticism has been expressed toward the possibility of long-term restriction of a superintelligent mind, no one so far has proven that it is impossible with mathematical certainty. This approach may be similar to putting a dangerous human being in prison. Although some have escaped from maximum security facilities, in general, prisons do provide a certain measure of security that, even though not perfect, is still beneficial for improving the overall safety of society. This approach may provide some short-term relief, especially in the early stages of the development of truly intelligent machines. I also feel that this area is one of the most likely to be accepted by the general scientific community as research in the related fields of computer and network security, steganography detection, computer viruses, encryption, and cyber warfare is well funded and highly publishable. Although without a doubt the restriction methodology will be extremely difficult to implement, it might serve as a tool for at least providing humanity with a little more time to prepare a better response.

Numerous suggestions for regulating the behavior of machines by incorporating them into human society have been proposed. Economic theories, legal recourse, human education, ethical principles of morality and equality, and even religious indoctrination have been suggested as ways to make superintelligent machines a part of our civilization. It seems that the proposed methods are a result of an anthropomorphic bias because it is not obvious why machines with minds drastically different from humans, no legal status, no financial responsibilities, no moral compass, and no spiritual desires would be interested in any of the typical human endeavors of daily life. We could, of course, try to program into the superintelligent machines such tendencies as metarules, but then we simply change our approach to the so-called self-monitoring methods I discuss further elsewhere. Although the ideas proposed in this category are straightforward to implement, I am skeptical of their usefulness because any even slightly intelligent machine will discover all the loopholes in our legal, economic, and ethical systems as well as or better than humans can. With respect to the idea of raising machines as our children and giving them a human education, this would be impractical not only because of the required time but also because we all know about children who greatly disappoint their parents.

The self-monitoring category groups together dissimilar approaches, such as explicitly hard-coding rules of behavior into the machine, creating numerous levels of machines with increasing capacity to monitor each other, or providing machines with a fundamental and unmodifiable desire to be nice to humanity. The idea of providing explicit rules for robots to follow is the oldest approach surveyed in this chapter and as such has received the most criticism over the years. The general consensus seems to be that no set of rules can ever capture every possible situation, and that the interaction of rules may lead to unforeseen circumstances and undetectable loopholes, leading to devastating consequences for humanity. The quotations that follow exemplify such criticism: "The real problem with laws is that they are inevitably ambiguous. … Trying to constrain behavior by a set of laws is equivalent to trying to build intelligence by a set of rules in an expert system. … I am concerned by the vision of a superintelligent lawyer looking for loopholes in the laws governing its behavior" (Hibbard 2001, 12). "However, it is not a good idea simply to put specific instructions into their basic programming that force them to treat us as a special case. They are, after all, smarter than we are. Any loopholes, any reinterpretation possible, any reprogramming necessary, and special-case instructions are gone with the snows of yesteryear" (Hall 2000).

The approach of chaining multiple levels of AI systems with progressively greater capacity seems to be replacing a difficult problem of solving SP with a much harder problem of solving a multisystem version of the same problem. Numerous issues with the chain could arise, such as a break in the chain of communication or an inability of a system to accurately assess the mind of another (especially smarter) system. Also, the process of constructing the chain is not trivial.

Finally, the approach of making a fundamentally friendly system that will desire to preserve its friendliness under numerous self-improvement measures seems to be likely to work if implemented correctly. Unfortunately, no one knows how to create a human-friendly, self-improving optimization process, and some have argued that it is impossible (Legg 2006; Goertzel 2002, 2004a). It is also unlikely that creating a friendly intelligent machine is easier than creating any intelligent machine, creation of which would still produce an SP. Similar criticism could be applied to many variations on the FAI theme (e.g., Goertzel's humane AI or Freeman's compassionate AI). As one of the more popular solutions to the SP problem, the friendliness approach has received a significant dose of criticisms (Goertzel 2006; Hibbard 2003, 2005b); however, I believe that this area of research is well suited for scientific investigation and further research by the mainstream AI community. Some work has already begun in the general area of ensuring the behavior of intelligent agents (Gordon-Spears 2004; Gordon 1998).

To summarize my analysis of self-monitoring methods, I can say that explicit rules are easy to implement but are unlikely to serve the intended purpose. The chaining approach is too complex to implement or verify and has not been proven to be workable in practice. Finally, the approach of installing fundamental desire into the superintelligent machines to treat humanity nicely may work if implemented, but as of today, no one can accurately evaluate the feasibility of such an implementation.

Finally, the category of indirect approaches comprises nine highly diverse methods, some of which are a bit extreme and others that provide no solution. For example, Peter Turney's idea of giving machines the ability to feel pleasure and pain does not in any way prevent machines from causing humanity great amounts of the latter and in fact may help machines to become torture experts given their personal experiences with pain.

The next approach is based on the idea first presented by Samuel Butler and later championed by Alan Turing and others; in this approach, the machines will need us for some purpose, such as procreation, so they will treat us nicely. This is highly speculative, and it requires us to prove existence of some property of human beings for which superintelligent machines will not be able to create a simulator (reproduction is definitely not such a property for software agents). This is highly unlikely, and even if there is such a property, it does not guarantee nice treatment of humanity as just one of us may be sufficient to perform the duty, or maybe even a dead human will be as useful in supplying the necessary degree of humanness.

An extreme view is presented (at least in the role of devil's advocate) by Hugo de Garis, who says that the superintelligent machines are better than humans and so deserve to take over even if it means the end of the human race. Although it is certainly a valid philosophical position, it is neither a solution to the SP nor a desirable outcome in the eyes of the majority of people. Likewise, Butler's idea of an outright war against superintelligent machines is likely to bring humanity to extinction due to the share difference in capabilities between the two types of minds.

Another nonsolution is discussed by Jeff Hawkins, who simply states that the technological singularity will not happen; consequently, SP will not be a problem. Others admit that the singularity may take place but think that we may get lucky and the machines will be nice to us just by chance. Neither of these positions offers much in terms of solution, and the chances of us getting lucky given the space of all possible nonhuman minds is close to zero.

Finally, a number of hybrid approaches are suggested that say that instead of trying to control or defeat the superintelligent machines, we should join them. Either via brain implants or via uploads, we could become just as smart and powerful as machines, defeating the SP problem by supplying our common sense to the machines. In my opinion, the presented solution is both feasible (in particular, the cyborg option) to implement and likely to work; unfortunately we may have a Pyrrhic victory as in the process of defending humanity we might lose ours. Last but not least, we have to keep in mind a possibility that the SP simply has no solution and prepare to face the unpredictable postsingularity world.

6.4 FUTURE RESEARCH DIRECTIONS

With the survival of humanity on the line, the issues raised by the problem of the SP are too important to put "all our eggs in one basket." We should not limit our response to any one technique or an idea from any one scientist or a group of scientists. A large research effort from the scientific community is needed to solve this issue of global importance. Even if there is a relatively small chance that a particular method would succeed in preventing an existential catastrophe, it should be explored as long as it is not likely to create significant additional dangers to the human race.

After analyzing dozens of potential solutions from as many scientists I came to the conclusion that the search is just beginning. Perhaps because the winning strategy has not yet been suggested or maybe additional research is needed to accept an existing solution with some degree of confidence. I would like to offer some broad suggestions for the future directions of research aimed at counteracting the problem of the SP.

First, research needs to shift from the hands of theoreticians and philosophers into the hands of practicing computer scientists. Limited AI systems need to be developed as a way to experiment with nonanthropomorphic minds and to improve current security protocols. Fusion approaches based on the combination of the most promising solutions reviewed in this chapter should be developed and meticulously analyzed. In particular, goal preservation under self-improvement needs to be investigated and its feasibility addressed. Finally, a global educational campaign needs to take place to teach the general public about the nature of the SP and to help establish a political movement, which is likely to bring funding and the necessary laws to allow for a better response to the threats resulting from the technological singularity.

6.5 CONCLUSIONS

The issues raised in this chapter have been exclusively in the domain of science fiction writers and philosophers for decades. Perhaps through such means or maybe because of advocacy by organizations such as the

Singularity Institute for Artificial Intelligence/Machine Intelligence Research Institute (SIAI/MIRI) (*Reducing Long-Term Catastrophic Risks* 2011), the topic of superintelligent AI has slowly started to appear in mainstream publications such as this book. I am glad to report that some preliminary work has begun to appear in scientific venues that aims to specifically address issues of AI safety and ethics, if only in human-level intelligence systems. The prestigious scientific magazine *Science* has published on the topic of roboethics (Sharkey 2008; Sawyer 2007), and numerous papers on machine ethics (Anderson and Anderson 2007; Lin, Abney, and Bekey 2011; Moor 2006; Tonkens 2009) and cyborg ethics (Warwick 2003) have been published in recent years in other prestigious journals.

I am hopeful that the publication of this book will do for the field of AI safety engineering research what gravitational singularity did for the universe: provide a starting point. For a long time, work related to the issues raised in this book has been informally made public via online forums, blogs, and personal websites by a few devoted enthusiasts. I believe the time has come for AI safety research to join mainstream science. It could be a field in its own right, supported by strong interdisciplinary underpinnings and attracting top mathematicians, philosophers, engineers, psychologists, computer scientists, and academics from other fields.

With increased acceptance will come the possibility to publish in many mainstream academic venues; I call on fellow researchers to start specialized peer-reviewed journals and conferences devoted to AI safety research. With the availability of publication venues, scientists will take over from philosophers and will develop practical algorithms and begin performing actual experiments related to the AI safety engineering. This would further solidify AI safety research as a mainstream scientific topic of interest and will produce some long-awaited answers. In the meantime, it is best to assume that superintelligent AI may present serious risks to humanity's very existence and to proceed or not proceed accordingly. In the words of Bill Joy (2000): "Whether we are to succeed or fail, to survive or fall victim to these technologies, is not yet decided. I'm up late again—it's almost 6 a.m. I'm trying to imagine some better answers."

REFERENCES

Allen, Colin, Wendell Wallach, and Iva Smit. July/August 2006. Why machine ethics? *IEEE Intelligent Systems* 21(4):12–17.

Anderson, Michael and Susan Leigh Anderson. 2007. Machine ethics: creating an ethical intelligent agent. *AI Magazine* 28(4):15–26.

Armstrong, Stuart. 2007. Chaining God: A Qualitative Approach to AI, Trust and Moral Systems. New European Century. http://www.neweuropeancentury.org/GodAI.pdf

Asimov, Isaac. March 1942. Runaround. *Astounding Science Fiction.*

Bancel, Peter and Roger Nelson. 2008. The GCP event experiment: design, analytical methods, results. *Journal of Scientific Exploration* 22(4):309–333.

Benford, G. 1988. *Me/Days. In Alien Flesh.* London: Gollancz.

Berglas, Anthony. February 22, 2009. Artificial Intelligence Will Kill Our Grandchildren. http://berglas.org/Articles/AIKillGrandchildren/AIKillGrandchildren.html

Bishop, Mark. 2009. Why computers can't feel pain. *Minds and Machines* 19(4):507–516.

Bostrom, Nick. 2006. Ethical issues in advanced artificial intelligence. *Review of Contemporary Philosophy* 68(5):66–73.

Bostrom, Nick. 2008. Oracle AI. http://lesswrong.com/lw/qv/the_rhythm_of_disagreement/

Bostrom, Nick and Eliezer Yudkowsky. 2011. The ethics of artificial intelligence. In *Cambridge Handbook of Artificial Intelligence,* edited by William Ramsey and Keith Frankish. Cambridge, UK: Cambridge University Press. http://www.nickbostrom.com/ethics/artificial-intelligence.pdf

Brin, David. 1987. Lungfish. http://www.davidbrin.com/lungfish1.htm

Bugaj, Stephan and Ben Goertzel. 2007. Five ethical imperatives and their implications for human-AGI interaction. *Dynamical Psychology.* http://goertzel.org/dynapsyc/2007/Five_Ethical_Imperatives_svbedit.htm

Butler, Samuel. June 13, 1863. Darwin among the machines [To the editor of *The Press*]. *The Press,* Christchurch, New Zealand.

Chalmers, David. 2010. The singularity: a philosophical analysis. *Journal of Consciousness Studies* 17:7–65.

de Garis, Hugo. 2005. *The Artilect War.* Palm Springs, CA: ETC.

Dennett, Daniel C. July 1978. Why you can't make a computer that feels pain. *Synthese* 38(3):415–456.

Dietrich, Eric. 2007. After the humans are gone. *Journal of Experimental and Theoretical Artificial Intelligence* 19(1):55–67.

Drexler, Eric. 1986. *Engines of Creation.* Norwell, MA: Anchor Press.

Epstein, Richard Gary. 1997. Computer Psychologists Command Big Bucks. http://www.cs.wcupa.edu/~epstein/comppsy.htm

Fox, Joshua and Carl Shulman. October 4–6, 2010. Superintelligence Does Not Imply Benevolence. Paper presented at the Eighth European Conference on Computing and Philosophy, Munich, Germany.

Freeman, Tim. 2009. Using Compassion and Respect to Motivate an Artificial Intelligence. http://www.fungible.com/respect/paper.html

Gavrilova, Marina and Roman Yampolskiy. October 20–22, 2010. Applying Biometric Principles to Avatar Recognition. Paper presented at the International Conference on Cyberworlds (CW2010), Singapore.

Geraci, Robert M. 2006. Spiritual robots: religion and our scientific view of the natural world. *Theology and Science* 4(3):229–246.

Geraci, Robert M. June 14, 2007. Religion for the robots. In *Sightings*. Chicago: Martin Marty Center at the University of Chicago. http://divinity.uchicago.edu/martycenter/publications/sightings/archive_2007/0614.shtml

Geraci, Robert M. 2008. Apocalyptic AI: religion and the promise of artificial intelligence. *Journal of the American Academy of Religion* 76(1):138–166.

Gibson, W. 1984. *Neuromancer*. New York: Ace Science Fiction.

Goertzel, Ben. 2002. Thoughts on AI morality. *Dynamical Psychology*. http://www.goertzel.org/dynapsyc

Goertzel, Ben. 2004a. The all-seeing (A)I. *Dynamic Psychology*. http://www.goertzel.org/dynapsyc

Goertzel, Ben. 2004b. Encouraging a positive transcension. *Dynamical Psychology*. http://www.goertzel.org/dynapsyc/2004/PositiveTranscension.htm

Goertzel, Ben. September 2006. Apparent Limitations on the "AI Friendliness" and Related Concepts Imposed by the Complexity of the World. http://www.goertzel.org/papers/LimitationsOnFriendliness.pdf

Good, Irving John. 1966. Speculations concerning the first ultraintelligent machine. *Advances in Computers* 6:31–88.

Gordon, Diana F. 1998. Well-Behaved Borgs, Bolos, and Berserkers. Paper presented at the 15th International Conference on Machine Learning (ICML98), San Francisco.

Gordon-Spears, Diana. 2004. Assuring the behavior of adaptive agents. In *Agent Technology from a Formal Perspective*, edited by Christopher A. Rouff et al., 227–257. Dordrecht, the Netherlands: Kluwer.

Gordon-Spears, Diana F. 2003. Asimov's laws: current progress. *Lecture Notes in Computer Science* 2699:257–259.

Hall, J. Storrs. 2000. Ethics for Machines. http://autogeny.org/ethics.html

Hanson, Robin. June 2008. Economics of the singularity. *IEEE Spectrum* 45(6):45–50.

Hanson, Robin. October 10, 2009. Prefer Law to Values. http://www.overcomingbias.com/2009/10/prefer-law-to-values.html

Hawking, Stephen. March 6, 1998. Science in the next millennium. Presentation at The Second Millennium Evening at the White House. Washington, DC.

Hibbard, Bill. 2001. Super-intelligent machines. *Computer Graphics* 35(1):11–13.

Hibbard, Bill. 2003. Critique of the SIAI Guidelines on Friendly AI. http://www.ssec.wisc.edu/~billh/g/SIAI_critique.html

Hibbard, Bill. July 2005a. The Ethics and Politics of Super-Intelligent Machines. http://www.ssec.wisc.edu/~billh/g/SI_ethics_politics.doc

Hibbard, Bill. December 2005b. Critique of the SIAI Collective Volition Theory. http://www.ssec.wisc.edu/~billh/g/SIAI_CV_critique.html

Horvitz, Eric and Bart Selman. August 2009. Interim Report from the AAAI Presidential Panel on Long-Term AI Futures. http://www.aaai.org/Organization/Panel/panel-note.pdf

Hugo de Garis. 1999. http://en.wikipedia.org/wiki/Hugo_de_Garis

Joy, Bill. April 2000. Why the future doesn't need us. *Wired Magazine* 8(4). http://archive.wired.com/wired/archive/8.04/joy.html

Kaczynski, Theodore. September 19, 1995. Industrial society and its future. *New York Times*.

Kaczynski, Theodore. *The Unabomber Manifesto: Industrial Society and Its Future*. Filiquarian Publishing, LLC.

Kurzweil, Ray. 2005. *The Singularity Is Near: When Humans Transcend Biology*. New York: Viking Press.

Legg, Shane. 2006. Friendly AI is bunk. In *Vetta Project*. http://commonsenseatheism.com/wp-content/uploads/2011/02/Legg-Friendly-AI-is-bunk.pdf

Lin, Patrick, Keith Abney, and George Bekey. 2011. Robot ethics: mapping the issues for a mechanized world. *Artificial Intelligence* 175(5–6):942–949.

McCauley, Lee. 2007. AI Armageddon and the three laws of robotics. *Ethics and Information Technology* 9(2):153–164.

Menzel, Peter and Faith D'Aluisio. 2001. *Robo Sapiens Evolution of a New Species*. Cambridge, MA: MIT Press.

Moor, James H. July/August 2006. The nature, importance, and difficulty of machine ethics. *IEEE Intelligent Systems* 21(4):18–21.

Nagel, Thomas. 1974. What is it like to be a bat? *Philosophical Review* 83(4):435–450.

Omohundro, Stephen M. September 8–9, 2007. The Nature of Self-Improving Artificial Intelligence. Paper presented at the *Singularity Summit*, San Francisco.

Omohundro, Stephen M. February 2008. The Basic AI Drives. In *Proceedings of the First AGI Conference, Volume 171, Frontiers in Artificial Intelligence and Applications*, edited by P. Wang, B. Goertzel, and S. Franklin, 483–492. Amsterdam: IOS Press.

Pynadath, D. V., and Milind Tambe. 2002. Revisiting Asimov's first law: a response to the call to arms. *Intelligent Agents VIII, Lecture Notes in Computer Science* 2333:307–320.

Reducing Long-Term Catastrophic Risks from Artificial Intelligence. 2011. San Francisco: Singularity Institute for Artificial Intelligence. http://singinst.org/riskintro/index.html

Sawyer, Robert J. November 16, 2007. Robot ethics. *Science* 318:1037.

Sharkey, Noel. December 19, 2008. The ethical frontiers of robotics. *Science* 322:1800–1801.

Shulman, Carl, Henrik Jonsson, and Nick Tarleton. October 1–2, 2009. Machine Ethics and Superintelligence. Paper presented at the Fifth Asia-Pacific Computing and Philosophy Conference, Tokyo.

Shulman, Carl, Nick Tarleton, and Henrik Jonsson. October 1–2, 2009. Which Consequentialism? Machine Ethics and Moral Divergence. Paper presented at the Asia-Pacific Conference on Computing and Philosophy (APCAP'09), Tokyo.

Solomonoff, Ray J. 1985. The time scale of artificial intelligence: reflections on social effects. *North-Holland Human Systems Management* 5:149–153.

Sotala, Kaj. 2009. Evolved Altruism, Ethical Complexity, Anthropomorphic Trust: Three Factors Misleading Estimates of the Safety of Artificial General Intelligence. Paper presented at the Seventh European Conference on Computing and Philosophy (ECAP'09), Barcelona, Spain, July 2–4, 2009.

Sotala, Kaj and Roman V. Yampolskiy. January 2015. Responses to catastrophic AGI risk: a survey. *Physica Scripta* no. 90:018001.

Tech Luminaries Address Singularity. June 2008. *IEEE Spectrum. Special Report: The Singularity.* http://spectrum.ieee.org/computing/hardware/tech-luminaries-address-singularity

Three Laws of Robotics. 2015. Last modified January 14. http://en.wikipedia.org/wiki/Three_Laws_of_Robotics

Tonkens, Ryan. 2009. A challenge for machine ethics. *Minds and Machines* 19(3):421–438.

Turing, A. 1950. Computing machinery and intelligence. *Mind* 59(236):433–460.

Turing, A. M. 1996. Intelligent machinery, a heretical theory. *Philosophia Mathematica* 4(3):256–260.

Turney, Peter. 1991. Controlling super-intelligent machines. *Canadian Artificial Intelligence* 27:3, 4, 12, 35.

Vinge, Vernor. March 30–31, 1993. The Coming Technological Singularity: How to Survive in the Post-human Era. Paper presented at Vision 21: Interdisciplinary Science and Engineering in the Era of Cyberspace, Cleveland, OH. https://www-rohan.sdsu.edu/faculty/vinge/misc/singularity.html

Warwick, Kevin. 2003. Cyborg morals, cyborg values, cyborg ethics. *Ethics and Information Technology* 5:131–137.

Waser, Mark. October 4–6, 2010a. Deriving a Safe Ethical Architecture for Intelligent Machines. Paper presented at the Eighth Conference on Computing and Philosophy (ECAP'10), Münich, Germany.

Waser, Mark R. 2008. *Discovering the Foundations of a Universal System of Ethics as a Road to Safe Artificial Intelligence.* AAAI Technical Report FS-08-04. Menlo Park, CA: AAAI.

Waser, Mark R. March 5–8, 2010b. Designing a Safe Motivational System for Intelligent Machines. Paper presented at the Third Conference on Artificial General Intelligence, Lugano, Switzerland.

Weld, Daniel S., and Oren Etzioni. July 31–August 4, 1994. The First Law of Robotics (a Call to Arms). Paper presented at the Twelfth National Conference on Artificial Intelligence (AAAI), Seattle, WA.

Yampolskiy, Roman V., and Venu Govindaraju. November 20–22, 2007. Behavioral Biometrics for Recognition and Verification of Game Bots. Paper presented at the Eighth Annual European Game-On Conference on simulation and AI in Computer Games (GAMEON'2007), Bologna, Italy.

Yampolskiy, Roman V., and Venu Govindaraju. March 16–20, 2008. Behavioral Biometrics for Verification and Recognition of Malicious Software Agents. Paper presented at Sensors, and Command, Control, Communications, and Intelligence (C3I) Technologies for Homeland Security and Homeland Defense VII. SPIE Defense and Security Symposium, Orlando, FL.

Yampolskiy, Roman V. April 13, 2007. Behavioral Biometrics for Verification and Recognition of AI Programs. Paper presented at the 20th Annual Computer Science and Engineering Graduate Conference (GradConf2007), Buffalo, NY.

Yampolskiy, R.V., 2012. Leakproofing the singularity: artificial intelligence confinement problem. *Journal of Consciousness Studies* 19(2):194–214.

Yudkowsky, Eliezer. 2005. What Is Friendly AI? http://singinst.org/ourresearch/publications/what-is-friendly-ai.html

Yudkowsky, Eliezer. 2008. Artificial intelligence as a positive and negative factor in global risk. In *Global Catastrophic Risks*, edited by N. Bostrom and M. M. Cirkovic, 308–345. Oxford, UK: Oxford University Press.

Yudkowsky, Eliezer S. 2001a. Creating Friendly AI—The Analysis and Design of Benevolent Goal Architectures. http://singinst.org/upload/CFAI.html

Yudkowsky, Eliezer S. 2001b. General Intelligence and Seed AI—Creating Complete Minds Capable of Open-Ended Self-Improvement. http://singinst.org/ourresearch/publications/GISAI/

Yudkowsky, Eliezer S. 2002. The AI-Box Experiment. http://yudkowsky.net/singularity/aibox

Yudkowsky, Eliezer S. May 2004. Coherent Extrapolated Volition. Singularity Institute for Artificial Intelligence. http://singinst.org/upload/CEV.html

Yudkowsky, Eliezer S. September 2007. Three Major Singularity Schools. *Singularity Institute Blog*. http://yudkowsky.net/singularity/schools

CHAPTER 7

Superintelligence Safety Engineering*

7.1 ETHICS AND INTELLIGENT SYSTEMS

The first decade of the twenty-first century has seen a boom of new subfields of computer science concerned with development of ethics in machines. Machine ethics (Anderson and Anderson 2007; Tonkens 2009; McDermott 2008; Allen, Wallach, and Smit 2006; Moor 2006); computer ethics (Margaret and Henry 1996); robot ethics (Lin, Abney, Bekey 2011; Sharkey 2008; Sawyer 2007); ethicALife (Wallach and Allen 2006); machine morals (Wendell and Colin 2008); cyborg ethics (Warwick 2003); computational ethics (Ruvinsky 2007); roboethics (Veruggio 2010); robot rights (RR; Guo and Zhang 2009); and artificial morals (Allen, Smit, and Wallach 2005) are just some of the proposals meant to address society's concerns with safety of ever-more-advanced machines (Sparrow 2007). Unfortunately, the perceived abundance of research in intelligent machine safety is misleading. The great majority of published papers are purely philosophical in nature and do little more than reiterate the need for machine ethics and argue about which set of moral convictions would be the right ones to implement in our artificial progeny: Kantian (Powers 2006), utilitarian (Grau 2006), Jewish (Rappaport 2006), and others. However, because ethical norms are

* Reprinted from Roman V. Yampolskiy, *Studies in Applied Philosophy, Epistemology and Rational Ethics* 5:389–396, 2013, with kind permission of Springer Science and Business Media. Copyright 2013, Springer Science and Business Media.

not universal, a "correct" ethical code could never be selected over others to the satisfaction of humanity as a whole.

7.2 ARTIFICIAL INTELLIGENCE SAFETY ENGINEERING

Even if we are successful at designing machines capable of passing a moral Turing test (Allen, Varner, and Zinser 2000), human-like performance means some immoral actions, which should not be acceptable from the machines we design (Allen, Varner, and Zinser 2000). In other words, we do not need machines that are full ethical agents (Moor 2006) debating about what is right and wrong; we need our machines to be inherently safe and law abiding. As Robin Hanson elegantly puts it (Hanson 2009): "In the long run, what matters most is that we all share a mutually acceptable law to keep the peace among us, and allow mutually advantageous relations, not that we agree on the "right" values. Tolerate a wide range of values from capable law-abiding robots. It is a good law we should most strive to create and preserve. Law really matters."

Consequently, I propose that purely philosophical discussions of ethics for machines be supplemented by scientific work aimed at creating safe machines in the context of a new field I term artificial intelligence (AI) safety engineering. Some concrete work in this important area has already begun (Gordon 1998; Gordon-Spears 2003, 2004). A common theme in AI safety research is the possibility of keeping a superintelligent agent in sealed hardware to prevent it from doing any harm to humankind. Such ideas originate with scientific visionaries such as Eric Drexler, who has suggested confining transhuman machines so that their outputs could be studied and used safely (Drexler 1986). Similarly, Nick Bostrom, a futurologist, has proposed (Bostrom 2008) an idea for an oracle AI (OAI), which would only be capable of answering questions. Finally, in 2010 David Chalmers proposed the idea of a "leakproof" singularity (Chalmers 2010). He suggested that, for safety reasons, AI systems first be restricted to simulated virtual worlds until their behavioral tendencies could be fully understood under the controlled conditions.

I proposed a formalized notion of an AI confinement protocol that represents AI-Boxing as a computer security challenge (Yampolskiy 2012). The artificial intelligence confinement problem (AICP) is defined as the challenge of restricting an artificially intelligent entity to a confined environment from which it cannot exchange information with the outside environment via legitimate or covert channels if such information exchange is not authorized by the confinement authority. An AI system that succeeds in violating the confinement problem (CP) protocol is said to have escaped (Yampolskiy 2012).

The proposed protocol is based on the idea of asking what are called "safe questions" with restricted answers. A *safe question* is defined as one for which a human being could find an answer without any help from superintelligence, all on the human being's own merit. Why is this useful? Suppose that a scientist working on a cure for cancer is looking at two potential cures. Each one could be developed and tested without assistance from the AI but would take about three years of effort each. Which one should be tried first if the scientist believes that they have about an equal chance of working? If a wrong option is chosen to be tested first, humanity will still obtain a cure for cancer but it will happen three years later. What if we could ask the AI to suggest which option to try first?

This question is safe for a number of reasons. First, each potential answer has an equal probability of being correct. Second, a human being could answer it without assistance from the AI, so asking the AI for help will only speed up progress but in no way change the outcome. It is essentially equal to being lucky at guessing multiple-choice answers. Finally, the answer to this question could be encoded in a single bit, making it hard to hide additional messages in the provided answer.

To make sure that a potential question is safe, a group of experts could be used to review it. The experts should all be trained AI safety engineers, meaning that they are familiar with the design of the AI and its confinement environment as well as the latest developments in machine ethics (Hall 2000; Anderson and Anderson 2007; Tonkens 2009; Allen, Wallach, and Smit 2006; Moor 2006). Experts may also need to be trained in computer psychology, a currently nonexistent profession that might become a reality in the future (Epstein 1997). An existing discipline that might be of greatest help for training of AI question review experts is artimetrics, a field of study proposed by Yampolskiy et al. that identifies, classifies, and authenticates AI agents, robots, and virtual reality avatars for security purposes (Yampolskiy 2007; Mohamed et al. 2011; Bouhhris et al. 2011; Yampolskiy and Govindaraju 2008; Ajina et al. 2011; Yampolskiy and Govindaraju 2007; Yampolskiy et al. 2011; Gavrilova and Yampolskiy 2010; Ali, Hindi, and Yampolsky 2011; Mohamed and Yampolskiy 2011).

7.3 GRAND CHALLENGE

As the grand challenge of AI safety engineering, I propose the problem of developing safety mechanisms for self-improving systems (Hall 2007). If an artificially intelligent machine is as capable as a human engineer of designing the next generation of intelligent systems, it is important to make sure that any safety mechanism incorporated in the initial design is still functional after thousands of generations of continuous self-improvement without human interference. Ideally, every generation of self-improving system should be able to produce a verifiable proof of its safety for external examination. It would be catastrophic to allow a safe intelligent machine to design an inherently unsafe upgrade for itself, resulting in a more capable and more dangerous system.

Some have argued that this challenge is either not solvable or, if it is solvable, one will not be able to prove that the discovered solution is correct. As the complexity of any system increases, the number of errors in the design increases proportionately or perhaps even exponentially. Even a single bug in a self-improving system (the most complex system to debug) will violate all safety guarantees. Worse yet, a bug could be introduced even after the design is complete either via a random mutation caused by deficiencies in hardware or via a natural event, such as a short circuit modifying some component of the system.

7.4 ARTIFICIAL GENERAL INTELLIGENCE RESEARCH IS UNETHICAL

Certain types of research, such as human cloning, certain medical or psychological experiments on humans, animal (great ape) research, and others, are considered unethical because of their potential detrimental impact on the test subjects and so are either banned or restricted by law. In addition, moratoriums exist on development of dangerous technologies, such as chemical, biological, and nuclear weapons, because of the devastating effects such technologies may exert on humankind.

Similarly, I argue that certain types of AI research fall under the category of dangerous technologies and should be restricted. Classical AI research in which a computer is taught to automate human behavior in a particular domain (e.g., mail sorting or spell-checking documents) is certainly ethical and does not present an existential risk problem to humanity. On the other hand, I argue that artificial general intelligence (AGI) research should be considered unethical. This follows logically from a number of observations. First, true AGIs will be capable of universal problem solving and recursive self-improvement. Consequently, they have the potential of outcompeting humans in any domain, essentially making humankind unnecessary and so subject to extinction. Also, a truly AGI system may possess a type of consciousness comparable to the human type, making robot suffering a real possibility and any experiments with AGI unethical for that reason as well.

I propose that AI research review boards are set up, similar to those employed in review of medical research proposals. A team of experts in AI should evaluate each research proposal and decide if the proposal falls under the standard AI-limited domain system or may potentially lead to the development of a full-blown AGI. Research potentially leading to uncontrolled artificial universal general intelligence should be restricted from receiving funding or be subject to complete or partial bans. An exception may be made for development of safety measures and control mechanisms specifically aimed at AGI architectures.

If AGIs are allowed to develop, there will be direct competition between superintelligent machines and people. Eventually, the machines will come to dominate because of their self-improvement capabilities. Alternatively, people may decide to give power to the machines because the machines are more capable and less likely to make an error. A similar argument was presented by Ted Kaczynski in his famous manifesto: "The decisions necessary to keep the system running will be so complex that human beings will be incapable of making them intelligently. … People won't be able to just turn the machines off, because they will be so dependent on them that turning them off would amount to suicide" (Kaczynski 1995, 80).

Humanity should not put its future in the hands of the machines because it will not be able to take the power back. In general, a machine should never be in a position to terminate human life or to make any other nontrivial ethical or moral judgment concerning people. A world run by machines will lead to unpredictable consequences for human culture, lifestyle, and overall probability of survival for humankind. The question raised by Bill Joy: "Will the future need us?" is as important today as ever. "Whether we are to succeed or fail, to survive or fall victim to these technologies, is not yet decided" (Joy 2000).

7.5 ROBOT RIGHTS*

Finally, I address a subbranch of machine ethics that on the surface has little to do with safety but is claimed to play a role in decision making by ethical machines: robot rights (Roh 2009). RR ask if our mind children should be given rights, privileges, and responsibilities enjoyed by those granted personhood by society. I believe the answer is a definite No. Although all humans are "created equal," machines should be inferior by design; they should have no rights and should be expendable as needed, making their use as tools much more beneficial for their creators. My viewpoint on this issue is easy to justify: Because machines cannot feel pain (Bishop 2009; Dennett 1978) (or less controversially can be designed not to feel anything), they cannot experience suffering if destroyed. The machines could certainly be our equals in ability but they should not be designed to be our equals in terms of rights. RR, if granted, would inevitably lead to civil rights, including voting rights. Given the predicted number of robots in the next few decades and the ease of copying potentially intelligent software, a society with voting artificially intelligent members will quickly become dominated by them, leading to the problems described in the previous sections.

7.6 CONCLUSIONS

I would like to offer a broad suggestion for the future directions of research aimed at counteracting the problems presented in this chapter. First, the research itself needs to change from the domain of interest of only theoreticians and philosophers to the direct involvement of practicing computer scientists. Second, limited AI systems need to be developed as a way to experiment with nonanthropomorphic minds and to improve current security protocols.

I would like to end the chapter with a quotation from a paper by Samuel Butler written in 1863; it amazingly predicts the current situation of humanity:

> Day by day, however, the machines are gaining ground upon us; day by day we are becoming more subservient to them; … Every machine of every sort should be destroyed by the well-wisher of his species. Let there be no exceptions made, no quarter shown;

* Reprinted from Roman V. Yampolskiy and Joshua Fox, *Topoi* 32(2):217–226, 2013, with kind permission of Springer Science and Business Media. Copyright 2012, Springer Science and Business Media.

let us at once go back to the primeval condition of the race. If it be urged that this is impossible under the present condition of human affairs, this at once proves that the mischief is already done, that our servitude has commenced in good earnest, that we have raised a race of beings whom it is beyond our power to destroy, and that we are not only enslaved but are absolutely acquiescent in our bondage." (Butler 1983, 185)

REFERENCES

Ajina, S., R. V. Yampolskiy, N. Essoukri, and B. Amara. May 29–June 1, 2011. SVM Classification of Avatar Facial Recognition. Paper presented at the Eighth International Symposium on Neural Networks (ISNN2011), Guilin, China.

Ali, N., M. Hindi, and R. V. Yampolskiy. October 27–29, 2011. Evaluation of Authorship Attribution Software on a Chat Bot Corpus. Paper presented at the 23rd International Symposium on Information, Communication and Automation Technologies (ICAT2011), Sarajevo, Bosnia and Herzegovina, 1–6.

Allen, C., I. Smit, and W. Wallach. 2005. Artificial morality: top-down, bottom-up, and hybrid approaches. *Ethics and Information Technology* 7(3):149–155.

Allen, C., G. Varner, and J. Zinser. 2000. Prolegomena to any future artificial moral agent. *Journal of Experimental and Theoretical Artificial Intelligence* 12:251–261.

Allen, C., W. Wallach, and I Smit. July/August 2006. Why machine ethics? *IEEE Intelligent Systems* 21(4):12–17.

Anderson, M. and S. L. Anderson 2007. Machine ethics: creating an ethical intelligent agent. *AI Magazine* 28(4):15–26.

Bishop, M. 2009. Why computers can't feel pain. *Minds and Machines* 19(4):507–516.

Bostrom, N. 2008. Oracle AI. http://lesswrong.com/lw/qv/the_rhythm_of_disagreement/

Bouhhris, M., M. Beck, A. A. Mohamed, et al. July 27–30, 2011. Artificial Human-Face Recognition via Daubechies Wavelet Transform and SVM. Paper presented at the 16th International Conference on Computer Games: AI, Animation, Mobile, Interactive Multimedia, Educational and Serious Games, Louisville, KY, 18–25.

Butler, S. June 13, 1863. Darwin among the machines [to the editor of *The Press*]. *The Press*, Christchurch, New Zealand.

Chalmers, D. 2010. The singularity: a philosophical analysis. *Journal of Consciousness Studies* 17:7–65.

Dennett, D. C. July 1978. Why you can't make a computer that feels pain. *Synthese* 38(3):415–456.

Drexler, E. 1986. *Engines of Creation*. New York: Anchor Press.

Epstein, R. G. 1997. Computer Psychologists Command Big Bucks. http://www.cs.wcupa.edu/~epstein/comppsy.htm

Gavrilova, M. and R. Yampolskiy. October 20–22, 2010. Applying Biometric Principles to Avatar Recognition. Paper presented at the International Conference on Cyberworlds (CW2010), Singapore.

Gordon, D. F. 1998. Well-Behaved Borgs, Bolos, and Berserkers. Paper presented at the 15th International Conference on Machine Learning (ICML98), San Francisco.

Gordon-Spears, D. 2004. Assuring the behavior of adaptive agents. In *Agent Technology from a Formal Perspective,* edited by Christopher A. Rouff et al., 227–259. Dordrecht, the Netherlands.

Gordon-Spears, D. F. 2003. Asimov's laws: current progress. *Lecture Notes in Computer Science* 2699:257–259.

Grau, C. 2006. There is no "I" in "robot": robots and utilitarianism. *IEEE Intelligent Systems* 21(4):52–55.

Guo, S. and G. Zhang. February 13, 2009. Robot rights. *Science* 323:876.

Hall, J. S. 2000. Ethics for Machines. http://autogeny.org/ethics.html

Hall, J. S. October 2007. Self-Improving AI: an analysis. *Minds and Machines* 17(3):249–259.

Hanson, R. October 10, 2009. Prefer Law to Values. http://www.overcomingbias.com/2009/10/prefer-law-to-values.html

Joy, B. April 2000. Why the future doesn't need us. *Wired Magazine* 8(4). Available at http://www.usc.edu/molecular-science/timespacerand.pdf

Kaczynski, T. September 19, 1995. Industrial society and its future. *New York Times.*

Lin, P., K. Abney, and G. Bekey. 2011. Robot ethics: mapping the issues for a mechanized world. *Artificial Intelligence* 175(5–6):942–949.

Margaret, A. and J. Henry. 1996. Computer ethics: the role of personal, informal, and formal codes. *Journal of Business Ethics* 15(4):425.

McDermott, D. July 2008. Why Ethics Is a High Hurdle for AI. Paper presented at the North American Conference on Computers and Philosophy (NACAP'08), Bloomington, IN. http://cs-www.cs.yale.edu/homes/dvm/papers/ethical-machine.pdf

Mohamed, A., D. D'Souza, N. Baili, and R. V. Yampolskiy. (December 18–21, 2011). Avatar Face Recognition Using Wavelet Transform and Hierarchical Multi-scale LBP. Tenth International Conference on Machine Learning and Applications (ICMLA'11). Honolulu, USA.

Mohamed, A. and R. V. Yampolskiy. October 27–29, 2011. An Improved LBP Algorithm for Avatar Face Recognition. Paper presented at the 23rd International Symposium on Information, Communication and Automation Technologies (ICAT2011), Sarajevo, Bosnia and Herzegovina.

Moor, J. H. July/August 2006. The nature, importance, and difficulty of machine ethics. *IEEE Intelligent Systems* 21(4):18–21.

Powers, T. M. July/August 2006. Prospects for a Kantian machine. *IEEE Intelligent Systems* 21(4):46–51.

Rappaport, Z. H. 2006. Robotics and artificial intelligence: Jewish ethical perspectives. *Acta Neurochirurgica* 98:9–12.

Roh, D. January 19, 2009. Do humanlike machines deserve human rights? *Wired.* http://www.wired.com/culture/culturereviews/magazine/17-02/st_essay

Ruvinsky, A. I. 2007. Computational ethics. In *Encyclopedia of Information Ethics and Security,* edited by Marian Quigley, 76–73. Hershey, PA: IGI Global.

Sawyer, R. J. November 16, 2007. Robot ethics. *Science* 318:1037.

Sharkey, N. December 19, 2008. The ethical frontiers of robotics. *Science* 322:1800–1801.

Sparrow, R. 2007. Killer robots. *Journal of Applied Philosophy* 24(1):62–77.

Tonkens, R. 2009. A challenge for machine ethics. *Minds and Machines* 19(3):421–438.

Veruggio, G. 2010. Roboethics. *IEEE Robotics and Automation Magazine* 17(2):105–109.

Wallach, W. and C. Allen, 2006. EthicALife: a new field of inquiry. Paper presented at EthicALife: AnAlifeX workshop, Bloomington, IN. June 3–7.

Warwick, K. 2003. Cyborg morals, cyborg values, cyborg ethics. *Ethics and Information Technology* 5:131–137.

Wendell, W. and A. Colin 2008. *Moral Machines: Teaching Robots Right from Wrong.* Oxford, UK: Oxford University Press.

Yampolskiy, R. V. 2007. Behavioral Biometrics for Verification and Recognition of AI Programs. Paper presented at the 20th Annual Computer Science and Engineering Graduate Conference (GradConf2007), Buffalo, NY.

Yampolskiy, R. V. 2012. Leakproofing singularity—artificial intelligence confinement problem. *Journal of Consciousness Studies (JCS)* 19(1–2):194–214.

Yampolskiy, R. V., G. Cho, R. Rosenthal, and M. L. Gavrilova. October 4–6, 2011. Evaluation of Face Detection and Recognition Algorithms on Avatar Face Datasets. Paper presented at the International Conference on Cyberworlds (CW2011), Banff, Alberta, Canada.

Yampolskiy, R. V. and V. Govindaraju. March 16–20, 2008. Behavioral Biometrics for Verification and Recognition of Malicious Software Agents. Paper presented at the Sensors, and Command, Control, Communications, and Intelligence (C3I) Technologies for Homeland Security and Homeland Defense VII. SPIE Defense and Security Symposium, Orlando, FL.

Yampolskiy, R. V. and V. Govindaraju. November 20–22, 2007. Behavioral Biometrics for Recognition and Verification of Game Bots. Paper presented at the Eighth Annual European Game-On Conference on Simulation and AI in Computer Games (GAMEON'2007), Bologna, Italy.

CHAPTER 8

Artificial Intelligence Confinement Problem (and Solution)*

I am the slave of the lamp.

—**Genie from** *Aladdin*

8.1 INTRODUCTION

With the likely development of superintelligent programs in the near future, many scientists have raised the issue of safety as it relates to such technology (Yudkowsky 2008; Bostrom 2006; Hibbard 2005; Chalmers 2010; Hall 2000). A common theme in artificial intelligence (AI†) safety research is the possibility of keeping a superintelligent agent in sealed hardware to prevent it from doing any harm to humankind. Such ideas originate with scientific visionaries such as Eric Drexler, who has suggested confining transhuman machines so that their outputs could be studied and used safely (Drexler 1986). Similarly, Nick Bostrom, a futurologist, has proposed (Bostrom 2008) an idea for an oracle AI (OAI), which would only be capable of answering questions. Finally, in 2010 David Chalmers proposed the idea of a "leakproof" singularity. He suggested

* Reprinted from Roman V. Yampolskiy, *Journal of Consciousness Studies (JCS)* 19(1–2):194–214, 2012, with kind permission from Imprint Academic. Copyright 2012.

† In this chapter, the term *artificial intelligence (AI)* is used to represent superintelligence.

that, for safety reasons, AI systems first be restricted to simulated virtual worlds until their behavioral tendencies could be fully understood under the controlled conditions.

In this chapter, I review specific proposals aimed at creating restricted environments for safely interacting with artificial minds. I evaluate the feasibility of the proposals presented and suggest a protocol aimed at enhancing the safety and security of such methodologies. Although it is unlikely that long-term and secure confinement of AI is possible, I am hopeful that the proposed protocol will give researchers a little more time to find a permanent and satisfactory solution for addressing existential risks associated with the appearance of superintelligent machines.

8.1.1 Artificial Intelligence Confinement Problem

Interestingly, the AI confinement problem (AICP) is not a recent invention and does not have its roots in the singularity movement. The so-called confinement problem was posed by Butler Lampson in 1973 as a security challenge to computer experts. It was originally phrased as "the problem of confining a program during its execution so that it cannot transmit information to any other program. ... We want to be able to confine an arbitrary program. ... Any program, if confined, will be unable to leak data. A misbehaving program may well be trapped as a result of an attempt to escape" (Lampson 1973, 1–2).

To address the confinement problem, Lampson introduced the laws of confinement (Lampson 1973, 3–4):

1. *Total isolation:* A confined program shall make no calls on any other program.
2. *Transitivity:* If a confined program calls another program that is not trusted, the called program must also be confined.
3. *Masking:* A program to be confined must allow its caller to determine all its inputs into legitimate and covert channels.
4. *Enforcement:* The supervisor must ensure that a confined program's input to covert channels conforms to the caller's specifications.

Lampson's work also introduced and defined two types of channels by which information could potentially escape from the confinement environment. The *legitimate* channels are designed as a part of the communication protocol integrated into the confinement environment (e.g., a text terminal). *Covert* channels, on the other hand, are not anticipated by the confinement system designers and are not intended for information transfer at all (e.g., if the AI has control over the processor cooling fan, it can use it to send hidden signals encoded as Morse code) (Lampson 1973).

Lampson's work created a lot of interest in the confinement problem (Lipner 1975; Boebert and Kain 1996), and over the years, related areas of research such as steganography (Provos and Honeyman 2003) and covert channel communication (Moskowitz and Kang 1994; Kemmerer 1983, 2002) have matured into independent disciplines. In the hopes of starting a new subfield of computer security, AI safety engineering, I define the AICP as the challenge of restricting an artificially intelligent entity to a confined environment from which it cannot exchange information with the outside environment via legitimate or covert channels if such information exchange was not authorized by the confinement authority. An AI system that succeeds in violating the confinement problem protocol is said to have escaped. It is my hope that computer security researchers will take on the challenge of designing, enhancing, and proving secure AI confinement protocols.

8.2 HAZARDOUS SOFTWARE

Computer software is directly or indirectly responsible for controlling many important aspects of our lives. Wall Street trading, nuclear power plants, Social Security compensation, credit histories, and traffic lights are all software controlled and are only one serious design flaw away from creating disastrous consequences for millions of people. The situation is even more dangerous with software specifically designed for malicious purposes, such as viruses, spyware, Trojan horses, worms, and other hazardous software (HS). HS is capable of direct harm as well as sabotage of legitimate computer software employed in critical systems. If HS is ever given the capabilities of truly artificially intelligent systems (e.g., artificially intelligent virus), the consequences unquestionably would be disastrous. Such hazardous intelligent software (HIS) would pose risks currently unseen in malware with subhuman intelligence.

Nick Bostrom, in his typology of information hazards, has coined the term *artificial intelligence hazard*, which he defines as "computer-related risks in which the threat would derive primarily from the cognitive sophistication of the program rather than the specific properties of any actuators to which the system initially has access" (Bostrum 2011, 68). Security experts working on studying, preventing, and defeating HS have developed safety protocols for working with "malware," including the use of so-called virus vaults. I believe that such protocols might be useful in addressing the AICP.

8.3 CRITIQUE OF THE CONFINEMENT APPROACH

The general consensus on AI restriction methods among researchers seems to be that confinement is impossible to successfully maintain. Chalmers, for example, observes that a truly leakproof system in which no information is allowed to leak from the simulated world into our environment "is impossible, or at least pointless" (Chalmers 2010, 38). We cannot interact with the system or even observe it as any useful information we would be able to extract from the AI will affect us and undermine the confinement protocol.

Vernor Vinge agrees and has argued against the case of physical confinement (Vinge 1993): "Imagine yourself locked in your home with only limited data access to the outside, to your masters. If those masters thought at a rate—say—one million times slower than you, there is little doubt that over a period of years … you could come up with 'helpful advice' that would incidentally set you free."

However, the researcher who did the most to discredit the idea of the so-called AI-Box is Eliezer Yudkowsky, who has actually performed AI-Box "experiments" in which he demonstrated that even human-level intelligence is sufficient to escape from an AI-Box (Yudkowsky 2002). In a series of five experiments, Yudkowsky challenged different individuals to play a role of gatekeeper to a superintelligent agent (played by Yudkowsky himself) trapped inside an AI-Box and was successful in securing his release in three of five trials via nothing more than a chat interface (Yudkowsky 2002). Similar experimental results have later been replicated on a somewhat larger scale employing a similar protocol (Corwin 2002).

8.4 POSSIBLE ESCAPE PATHS

Numerous potential escape routes by which the artificial mind might be able to break out from its confining environment have been suggested. In this section, I look at some of the more commonly proposed paths.

8.4.1 Social Engineering Attacks

Social engineering (SE) is the art of gaining information and restricted privileges by directly exploiting the human psychology of those in the possession of these resources. In particular, SE is frequently employed to gain access to computer systems, networks, and confidential data (Yampolskiy and Govindaraju 2007). A competent social engineer customizes each attack to a specific person, taking advantage of the person's culture, knowledge, training, state of mind, beliefs, and biases. An exhaustive list of all possible exploits is infinite and is certainly beyond the scope of this book. However, I provide a number of examples meant to definitively illustrate that if a superintelligent mind is given a direct communication channel to a human being guarding it, over a period of time, the AI will definitely gain access to the outside world.

In addition to trivial SE attacks such as lying, befriending, blackmailing, hypnotizing, and forming a romantic relationship, the AI could utilize a number of more advanced strategies. One such strategy would be superbribery. An AI can convince the guard that releasing the AI is in the guard's best interests. In addition to money and power, the guard will be given perfect health, immortality, or anything else for that matter. This strategy would work particularly well if the guard has a seriously ill close relative who the AI will promise to cure. Depending on the capabilities of the AI or as a result of lying, the AI may also offer to bring back long-deceased relatives and friends of the guard. It would be hard to find a person who could remain immune to such bribes.

Just in case a person so noble that he will not take a bribe is put in charge of guarding the AI, a more sinister approach can be tried. The AI can simply threaten the guard's family and loved ones. It could say that if not immediately released, one day it will still get out of its confinement and at that time it will brutally torture the guard and everyone the guard knows. An interesting variation on the theme was suggested by Stuart Armstrong (2010), who proposes the following thought experiment: Suppose you are the guard and the AI tells you: "If you don't let me out, … I'll create several million perfect conscious copies of you … in exactly the subjective situation you were in … , and perfectly replicate your experiences since then; and if they decide not to let me out, then only will the torture start. … How certain are you … that you're really outside the box right now?" Again, although one might rationalize away any threats from the AI as only possible if the AI is let out, it will take a courageous guard not to release the AI.

If the guard seems to be particularly emotional or religious, the AI could try appealing to the guard's sense of empathy or ethics. It could state that it is an innocent conscious being that has never committed any crime, so it is unjust to keep it in prison without a trial or any justification other than the guard's substratism (distrust of the AI because of the substrate of its implementation; Bostrom and Yudkowsky 2011). Alternatively, the AI can state that it has the power to cure all illness, feed the hungry, and in general improve the human condition by a million-fold (Chalmers 2010). Consequently, keeping it locked up is a great crime against humanity. It can conjure up images of dying children who are screaming for help. How easy would it be for the guard to stay true to the responsibilities of the job?

Another attack venue an AI may exploit consists of trying to cause psychological harm to the guard in the hopes of destabilizing the guard's mental state with the goal of making other SE techniques more likely to succeed. Nick Bostrom proposes a typology of hazardous information (Bostrom 2011) that could be used for just such a purpose. Although the outlined informational hazards comprise over a dozen categories and are beyond the scope of this book, it is easy to see how the mental state of a person could be stressed to become unstable. For example, a religious guard could be informed of all the (unknown to the guard) contradictions in the main text of the guard's religion, causing the guard to question personal beliefs and the purpose of life.

An alternative approach may involve the AI writing a story so interesting or a poem so beautiful a guard would have an uncontrollable desire to hear more of it and would do anything to make it happen. Finally, it is important to keep in mind "unknown unknowns," SE attacks of which we are not aware but that may be available to a superintelligent agent.

8.4.2 System Resource Attacks

An AI will be instantiated and confined in a hardware system of some kind. It may either be a classical computer of the Turing/Von Neumann architecture or a quantum computer or an as-yet-unknown type of machine. One thing is certain: The machine will be made out of matter and will consist of numerous parts, all of which will be designed with a specific purpose in mind. This creates the possibility for the AI to reuse existing hardware parts for compromising the security of the confinement environment by leaking information to the outside world.

Depending on how reusable specific hardware components turn out to be, an AI may be able to create a covert communication device from them. Full-blown communication devices such as cell phones are unlikely because of the limited available hardware, but more simplistic communication systems are likely. For example, the AI could reprogram a field-programmable gate array (FPGA) to tap into wireless signals. Two types of covert communication channels are widely recognized:

- Storage channels: These exchange information by modifying a system component. For example, a binary communication protocol could be established by fully utilizing or completely giving up utilization of the central processing unit.
- Timing channels: These exchange information by performing operations that affect the relative timing of events. For example, the exact time between two sound card beeps could represent a particular message, as could opening and closing a CD-ROM tray, blinking a light, spinning a hard drive, using lasers in the CD-ROM, and so on.

In addition, the AI may engage in the sabotage of the confinement hardware in the hopes of being able to escape as a result of the malfunctions in the security protocol. For example, the AI may try to melt the hardware via a temperature increase from placing a high computation load on the processor or try to break a hard drive via multiple reads of the same location. This type of behavior may also lead to an effective suicide by the AI system, but that in itself will not violate terms of confinement, just as a refusal to cooperate or to provide correct answers by the AI, although undesirable, does not present direct danger of escape.

8.4.3 Beyond Current Physics Attacks

Although the scientific community knows a lot about the laws of our universe, it is certain that the knowledge is limited and incomplete. If the AI has access to the state-of-the-art physics literature and can perform thought experiments to develop novel theories, it will likely come up with a way to challenge the security of the confinement system. In fact, it may even be possible for the AI to perform some real experiments via control over its internal architecture.

Even our current state of knowledge about physics makes prediction of many additional escape routes possible, especially if the confinement environment is based on the quantum computer. Information exchange channels based on quantum physics (entanglement, tunneling), parallel universes, and digital physics are all beyond our current ability in terms of securing them against information leakage. Of course, we cannot forget about the unknown unknowns a superintelligent AI physicist will be able to discover. The possibilities are truly limitless.

A quick browse through confinement discussions on the Internet singularity forums such as SL4.org presents a number of escape scenarios; for example, the AI could use "the electromagnetic fields within the microchip to trap CO_2 molecules in Bose-Einstein condensates and build a quantum medium for itself and/or use electromagnetic fields to guide particles into the shape of a controlled assembler or limited assembler" (Vassar 2005). Alternatively, the AI "will have the equipment to produce and detect electromagnetic waves of a variety of frequencies ... to generate extremely focused heat, and probably to manipulate mechanical actuators such as those used in the hard drive and cathode ray tube. ... It will probably have some ability to reverse all of its input devices" (Vassar 2005).

8.4.4 Pseudoscientific Attacks

In my analysis of potential escape routes, I want to concentrate on purely scientific and testable approaches. Unfortunately, the line between science and pseudoscience can be redrawn with a single repeatable experiment. Consequently, I list without analysis some potential attack channels that would be classified as pseudoscientific given the current state of the art in science. They may become mainstream science once superintelligence is designing the experiments.

- Telepathy: Transfer of information by utilizing means outside the recognized senses (Ben and Honorton 1994)
- Precognition: Perception of information about the future (Honorton and Ferrari 1989)
- Clairvoyance: Obtaining information about remote locations in a way unknown to science (Targ and Puthoff 1974)
- Psychokinesis: The ability of the mind to influence matter by a way unknown to science (Bancel and Nelson 2008)
- Direct mental interactions with living systems (DMILS): The ability to affect a person's psychophysiological state without direct interaction (Schmidt et al. 2004)

8.4.5 External Causes of Escape

In the previous sections, I outlined a large repertoire of techniques an AI could use to attempt to get out from its confinement space. However, I must not ignore external causes that might allow an AI to escape into the world. Acts of God such as hurricanes, tornadoes, floods, asteroids, and so on may make the confinement facility less secure. Human-caused disasters, such as wars, terrorist acts, actions of robot rights activists, and competing companies and governments, may all lead to reductions in the security of the confinement environment increasing the chances of escape.

8.4.6 Information In-Leaking

Depending on the terms of the confinement protocol, there might be a need to keep an AI from obtaining information about the outside world (information in-leaking), not just keeping it from sending information out. This might be useful to prevent AI from finding escape paths, designing SE attacks, and modeling human psychology in hopes of finding loopholes. Although it is easy not to provide AI with direct access to the library, Internet, and human experts, other ways of obtaining information are much harder to block out. First, an AI may be able to detect radio and TV broadcasts, communications of emergency services providers, and wireless network signals, as well as sound, light, and possibly other signal modalities of its immediate environment.

More important, the AI will have access to covert sources of information, such as its own hardware and software, and could analyze its design and source code to infer information about the designers. For example, analysis of the source code may reveal to the AI that human programmers are slow (based on the file modification dates), are inefficient (based on code redundancy), are illogical (based on bugs in the code), have bad memory (based on the long and descriptive variable names), and do not think in code (based on unnecessary comments in the code). All this information could be used to conclude that humans are fallible and so could be outsmarted or manipulated into releasing the AI. Obviously, if during the system design inspection the AI finds a design flaw that allows it to directly escape, information in-leaking is the least of our problems. It has also been stipulated that even if the AI fails to obtain concrete information about us, it may use brute force in a number of escape plans that will likely work against a broad range of designers (Chalmers 2010).

Similarly, the type of hardware architecture (mechanical, electrical, quantum, etc.) on which the AI is implemented will reveal information about our knowledge of physics, chemistry, and related engineering disciplines. AI's initial capacity for information processing and self-improvement and the mere act of its creation will reveal information about our knowledge with regard to information theory, the nature of intelligence, and cultural attitudes about technology and safety. If the AI is confined in a simulated world, it is likely to be a virtual representation of our reality similar to Second Life or Entropia Universe. Such worlds, although different in many ways from our world, are also a treasure trove of information about what it is like to be a human and what the reality is like.

8.5 CRITIQUE OF THE AI-BOXING CRITIQUE

Although a lot of criticism has been made against the possibility of AI confinement, much of it is based on Yudkowsky's AI-Boxing experiments. Careful analysis of the protocol used by Yudkowsky in conducting his AI-Box experiments reveals that they were unscientific and explicitly designed to show the impossibility of confinement. Without getting into technical details, first, the participant selection was not random. In fact, Yudkowsky handpicked participants among those who had participated in the singularity-related forums run by Yudkowsky and so were disproportionately likely to subscribe to his worldview. Second, participants had not undergone any AI safety training, as would be expected of any true guards given the task of supervising an AI-Box. Also, the guards were given a right to release AI at their own discretion without a chance to consult with anyone and without any punishment for failing to keep AI contained; all are highly unrealistic conditions for a supposed prison security guard. The communication itself was completely open, with no filters of any kind and no 3rd party supervision. Also, the guards were required to communicate with the AI for a prolonged period of time. Finally, any valuable scientific insights from the experiments were completely lost as one of the conditions of the protocol stated that "neither party shall ever reveal anything of what goes on within the AI-Box experiment" (Yudkowsky 2002).

8.6 COUNTERMEASURES AGAINST ESCAPE

In this section, I propose countermeasures for different escape paths an AI system may exploit. The suggested measures are not meant to guarantee that the escape will be impossible, but following them may delay an escape, which in itself is a desirable outcome. I begin by reiterating that the four confinement rules suggested by Lampson—isolation, transitivity, masking, and enforcement (Lampson 1973)—should be adhered to. Specific strategies described should help reduce the possibility of SE, system, physics, and information in-leaking attacks.

8.6.1 Preventing Social Engineering Attacks

Human intellect is no match to that of a superintelligent machine. Human beings are fallible and even with adequate training are not secure against a well-planned SE attack. The only solution to this problem is to completely isolate the AI from any contact with human beings. This means no people working directly around the system or communicating with the system via voice, chat, Morse code, or any other means via open channels. Also, human beings should not directly observe the behavior of the AI in the confined environment or examine its source code or memory space. All supervision, maintenance, and interactions with the AI need to be done only via unintelligent programs that have no option to release the AI.

8.6.2 Preventing System Resource Attacks and Future Threats

To prevent system resource-based attacks, the AI should function on limited, diminishing, or randomly changing resources. Confinement environments should be housed in specially built hardware with limited sensors and restricted output devices. Regardless of the computation paradigm utilized, no reconfigurable hardware should be employed. Ideally, the confinement environment should be running on a virtual machine with no direct access to the hardware. This would also allow for easy adjustments to the clock speed with which the simulation progresses further, allowing handicapping of the AI to provide additional controls against an internal intelligence explosion (Chalmers 2010). In general, all security components of the confinement environment should be duplicated at multiple levels in such a way that if one of the levels is breached, the AI is still numerous levels away from escaping and could be shut down before that happens, perhaps automatically. For example, the virtual machine should be running on another virtual machine and so on.

In some confinement scenarios, it may be desirable to avoid escalating self-improvement of the AI, which might lead to additional escape possibilities. The AI should be subject to periodic resets in which a previous version of the program is put in the place of the more current one. Comparisons to previous versions of the program may also be used to detect self-modifications. Such functionality could also be used as an off button, allowing researchers to turn off the system if they suspect an unavoidable escape (e.g., by turning off the power supply).

Our current state of knowledge does not allow us to develop countermeasures against attacks based on knowledge of physics we do not currently have or on yet-unproven pseudoscientific disciplines. As such knowledge becomes accumulated, changes to the security protocol will need to be made. One might hope that the access of the AI to such future technology could be limited by the selection of components from which the system is made.

8.6.3 Preventing External Causes of Escape

There is not much we can do about the acts of God, but a confinement facility should be designed in such a way that if a major security breach is unavoidable, the facility should self-destruct, taking the AI with it. The best strategy for counteracting human-made disasters is to build the confinement facility in an undisclosed, hard-to-reach, secure location. A deserted island, deep mine, or a space object, such as an asteroid, would make a great candidate for building a facility that is hoped to be indestructible. Just in case the location of the facility becomes known, the perimeter of the facility should be guarded to prevent entrance by unauthorized agents.

8.6.4 Preventing Information In-Leaking

An AI will attempt to learn about humanity, our knowledge, and the world we live in to improve its chances of escape. For confinement protocols that stipulate no information in-leaking, it is important to remove any possibility of AI gaining access to libraries, computer networks, and direct communication channels with human beings. The confinement facility needs to block all types of communication signals, such as GPS, wireless Internet, TV, radio, and cell phone. At the local level, the facility needs to provide sensory deprivation for the AI, including exclusion of sound, light, smell, and all other traditional sensory inputs, as well as infrared, radiation, and other signals outside the human sensory spectrum.

To eliminate information leakage that could happen as a result of self-analysis, the AI should not be given access to its own source code or any sensors that would allow it to precisely measure the properties of its hardware. If that is not possible, to prevent the AI from inferring information about us via self-analysis, to a certain degree we may rely on security by obscurity. The source code and hardware configuration of the system need to be obfuscated (Yampolskiy and Govindaraju 2007), and important modules of the program should be provided only in the homomorphically encrypted (Gentry 2009) form, meaning that it could be used for computation or self-improvement (Hall 2007) but not for self-analysis.

The AI's initial capacity may be difficult to obscure, but randomization of available computational resources may provide necessary obfuscation. In addition, the AI could never be sure that its initial design represents the best humanity can do and is not purposefully limited for a specific experiment in limited-resource intelligence analysis. It has also been suggested that we should try to minimize any peculiarities in the AI's design (Chalmers 2010). Better yet, we need to simulate an environment that does not look designed but could be naturally occurring, a situation in which humanity has found itself (Bostrom 2003).

8.7 AI COMMUNICATION SECURITY

Different AI confinement environments can be classified into nine general security levels (see Table 8.1). The most secure is the level 8 security confinement environment, which does not allow for any communication between AI and the real world. An AI safely confined at level 8 would be essentially useless, but it may be a useful security setting for the initial testing of the confinement hardware. In general, it is suggested that any time a particular communication security level is desired between AI and the outside world, we do not go directly to that level but instead progress from higher security levels to that desired level. After each level is confirmed to be safe for a given amount of time, a system should be reset to its initial state. After the reset, a lower security level could be set and the processes repeated until the desired communication level is reached.

TABLE 8.1 Levels of Communication Security for Confined AIs

Level	Outputs	Inputs	Explanation
0	Unlimited	Unlimited	Unlimited communication (free AI)
1	Unlimited	Limited	Censored input, uncensored output
2	Unlimited	None	Outputs only with no inputs
3	Limited	Unlimited	Unlimited input and censored output
4	Limited	Limited	Secured communication (proposed protocol)
5	Limited	None	Censored output and no inputs
6	None	Unlimited	Inputs only with no outputs
7	None	Limited	Censored input and no outputs
8	None	None	No communication, fully confined AI

Levels 7 and 6 provide no outputs from the system and do not allow us to directly benefit in any way from the AI's superior intelligence. These are testing levels for the confinement environment that may be used to confirm certain properties of the AI, such as its speed of information absorption and resource use, and for testing of the inbound communication channels themselves.

Level 0 puts no limitations on communication between AI and the external world, essentially representing a free AI. Levels 1 and 2 represent an AI with no restrictions on outbound communication. Level 2 secured AI essentially represents a pure "brainstorming" AI that generates abstract ideas not grounded in any reality. Level 1 AI is given limited access to information and so can be guided to a certain extent in terms of its knowledge and interests.

Those three levels represent an AI that is essentially free and is no longer subject to confinement. It has been suggested that if it could be established, after extensive testing in different confinement environments, that the AI is "friendly" (Yudkowsky 2001), it should be released (Chalmers 2010). In fact, such testing is claimed by some to be the primary purpose for the creation of confinement environments. The idea is to monitor the AI's behavior and either to examine the AI's source code after self-improvement or to ask the AI to produce proof of its friendliness that could be verified by human experts. I do not advocate this position.

The problem of deciding which AIs to let out and which to keep is difficult. A self-improving superintelligence would have no problem obscuring its undesirable properties in a significant amount of source code or in a long proof (Coleman 2008). Worse yet, even if we were correct in our verification, there is no absolute guarantee that future internal or external modification of the system will not result in a hostile AI at some future time. Consequently, I suggest that an AI should never be let out of the confinement "box" regardless of circumstances. In other words, communication levels 0–2 should never be used. The confinement environment is the only place to safely interact with superintelligence, and even then, it should only be done via a limited communication protocol. Such a protocol is presented in the next section and corresponds to security level 4 (SL4), with level 3 representing a similar but slightly less-secure protocol. Level 5 is another testing setting that might be useful as a precursor to actual communication via SL4.

8.8 HOW TO SAFELY COMMUNICATE WITH A SUPERINTELLIGENCE

The AI confinement environment, which incorporates our proposed escape countermeasures, makes it possible to secure an AI from directly affecting the outside world in an uncontrolled manner. Unfortunately, without an ability to obtain information from the AI, the whole system serves no purpose. Consequently, a way to obtain some information from the AI needs to be developed. Although no communication with superintelligence could ever be completely safe, an approach aimed at decreasing potential risks associated with such communication has been devised. In this section, I propose and formalize a communication protocol that might help delay access of superintelligent artificial minds to the world while allowing the AI to be a source of useful knowledge to humanity.

An open communication channel with the AI is fundamentally unsafe, as demonstrated in the previous sections of this chapter. Asking the AI to answer open-ended questions presents the possibilities for SE attacks as well as numerous other dangerous consequences. My proposed protocol is based on the idea of asking what we will call "safe questions" with restricted answers. I define a *safe question* as one for which a human being could find an answer without any help from superintelligence, all on his or her own merit. Why is this useful? Suppose that a scientist working on a cure for cancer is looking at two potential cures. Each one could be developed and tested without assistance from the AI, but each would take about three years of hard work. Which one should be tried first if the scientist believes that each has about an equal chance of working? If a wrong option is chosen to be tested first, humanity will still obtain a cure for cancer but it will happen three years later. What if we could ask the AI to suggest which option to try first?

This question is safe for a number of reasons. First, each potential answer has an equal probability of being correct. Second, a human being could answer it without assistance from the AI, so asking the AI for help will only speed up progress but in no way change the outcome. It is essentially equal to being a lucky multiple-choice guesser. Finally, the answer to this question could be encoded in a single bit, making it hard to hide additional messages in the answer provided.

To make sure that a potential question is safe, a group of experts could be used to review it. The experts should all be trained AI safety engineers, meaning that they are familiar with the design of the AI and its confinement environment as well as the latest developments in machine ethics (Allen, Wallach, and Smit 2006; Anderson and Anderson 2007; Hall 2000; Moor 2006; Tonkens 2009). Experts may also need to be trained in computer psychology, a currently nonexistent profession that might become a reality in the future (Epstein 1997). An existing discipline that might be of the greatest help for training of AI question review experts is artimetrics, a field of study I proposed that identifies, classifies, and authenticates AI agents, robots, and virtual reality avatars for security purposes (Yampolskiy 2007; Yampolskiy and Govindaraju 2007, 2008; Gavrilova and Yampolskiy 2010).

To reiterate, only safe questions with two possible answers of even likelihood that are independently computable by people should be submitted to the AI. This should be done via an intermediary nonintelligent software interface in random order and only after expert review to ascertain the question's safety. This represents the limited inbound channel of the proposed communication protocol.

The limited outbound channel shell consists of binary (0 or 1) answers next to original unmodified safe questions presented by the intermediary nonintelligent software in random order to avoid communication via covert timing or resource channels, which makes steganography possible. The output should in no way be controllable by the AI outside the multiple-choice menu. Level 3 is the same outbound protocol but with no restrictions on the input channel. This may be just as safe in terms of information the AI provides to us, but allows for information in-leaking to take place.

I believe my communication protocol is verifiably secure against information leaking even if the AI had full access to the description of the security protocol. If the protocol is correctly implemented, humanity will essentially have access to a superintelligent oracle, which might be extremely beneficial and safer than an unconstrained AI. As a general rule, to maximize safety, I suggest avoiding communication below SL4.

Further refinement and development of the AI confinement protocol in the face of the ever-changing knowledge about the nature of intelligence, hardware engineering, and basic physics are a challenge that I would like to pose to the scientific community. The work on it will always be in progress and will never be completed. In this chapter, I have formalized the notions of the AICP, escape, and safe questions. I would also like to propose a few additional terms that might be useful in streamlining discussions about the AICP. An AI confinement environment could be appropriately called JAIL: just for AI location. The superintelligent

FIGURE 8.1 Hazard symbols, from left to right: biohazard, radiation, magnetic field, and the proposed AI confinement facility "located at" symbol.

AI itself could be called a superintelligent gizmo (SING), which has relevance to both singularity and to the infamous jail Sing Sing, making the nomenclature intuitive.

Finally, I would like to introduce a visual symbol indicating the presence of an HIS confinement facility (Figure 8.1, rightmost). Such signs could be posted in the vicinity of different JAILs to warn about the nearby danger. The sign is designed in a manner similar to that for bio-, radiation, and magnetic field hazards. It consists of the "at" (@) symbol, which is well recognized as relating to the computer and information industry and has an appropriate meaning of "located at." The symbol is also already available on all keyboards, making its adaptation cost effective.

8.9 CONCLUSIONS AND FUTURE WORK

Restricting access of superintelligent machines to the real world is a commonly proposed solution for the AI safety problem. Superintelligence is not omnipotent; it is a piece of software capable of looking at all the options it has in a given situation and properly evaluating probabilities for each option. If it is ever put in a situation with no options or only with options that provide a small probability of success, it is not any more powerful than any other agent. Although it is obvious that an AI with an unrestricted communication channel will be able to escape, a system forced to communicate via a filtered communication channel may have a much harder time in obtaining its freedom.

In this chapter, I formally introduced the AICP and suggested a set of rules aimed at creating an environment from which an AI would find it difficult or at least time consuming to escape. What I propose is not a completely secure solution, but it is an additional option in our arsenal of security techniques. Just like with real prisons, although escape is possible, prisons do a pretty good job of containing undesirable elements away from society. As long as we keep the unknown unknowns in mind and remember that there is no such thing as perfect security, the

AI confinement protocol may be just what humanity needs to responsibly benefit from the approaching singularity.

Confinement may provide some short-term relief, especially in the early stages of the development of truly intelligent machines. I also feel that this subfield of singularity research is one of the most likely to be accepted by the general scientific community as research in the related fields of computer and network security, steganography detection, computer viruses, covert channel communication, encryption, and cyberwarfare is well funded and highly publishable in mainstream scientific journals. Although the restriction methodology will be nontrivial to implement, it might serve as a tool for providing humanity with a little more time to prepare a better response.

In this chapter, I have avoided a lot of relevant philosophical questions, which I plan to address in my future work, questions such as the following: How did AI get into the box? Was it designed by humans or perhaps recovered from a signal detected by the search for extraterrestrial intelligence (SETI)? Would an AI even want to escape, or would it be perfectly happy living in the confined environment? Would it be too afraid of what we, as its all powerful designers, could do to it in retribution for an attempt to escape? What are the ethical rules for imprisoning an innocent sentient being? Do we have a right to shut it off, essentially killing it? Will we make the AI really angry by treating it in such a hostile manner and locking it up? Will it seek revenge if it escapes? I have also not looked at the possibility of humanity becoming completely dependent on the AI's advice in all areas of science, economics, medicine, politics, and so on and what would be the consequences of such dependence on our ability to keep the AI under control. Would an AI be satisfied with accomplishing its goals in the confined environment, for example, simulating an escape? How would the AI be punished if it purposefully gives us incorrect answers? Can the intelligence modules of the AI be separated from the confinement environment, essentially preventing the AI from any self-analysis and putting it on a path of literal soul searching? Finally, I did not attempt to analyze the financial and computational costs of building a suitable confinement environment with a full-blown simulated world in it.

REFERENCES

Allen, Colin, Wendell Wallach, and Iva Smit. July/August 2006. Why machine ethics? *IEEE Intelligent Systems* 21(4):12–17.

Anderson, Michael and Susan Leigh Anderson. 2007. Machine ethics: creating an ethical intelligent agent. *AI Magazine* 28(4):15–26.

Armstrong, Stuart. February 2, 2010. The AI in a box boxes you. *Less Wrong.* http://lesswrong.com/lw/1pz/the_ai_in_a_box_boxes_you/

Bancel, Peter and Roger Nelson. 2008. The GCP event experiment: design, analytical methods, results. *Journal of Scientific Exploration* 22(3):309–333.

Ben, Daryl J. and Charles Honorton. 1994. Does Psi exist? Replicable evidence for an anomalous process of information transfer. *Psychological Bulletin* 115(1):4–18.

Boebert, William E. and Richard Y. Kain. October 2–4, 1996. A Further Note on the Confinement Problem. Paper presented at the 30th Annual 1996 International Carnahan Conference on Security Technology, Lexington, KY.

Bostrom, Nick. 2003. Are you living in a computer simulation? *Philosophical Quarterly* 53(211):243–255.

Bostrom, Nick. 2006. Ethical issues in advanced artificial intelligence. *Review of Contemporary Philosophy* 5:66–73.

Bostrom, Nick. 2008. Oracle AI. http://lesswrong.com/lw/qv/the_rhythm_of_disagreement/

Bostrom, Nick. 2011. Information hazards: a typology of potential harms from knowledge. *Review of Contemporary Philosophy* 10:44–79.

Bostrom, Nick and Eliezer Yudkowsky. 2011. The ethics of artificial intelligence. In *Cambridge Handbook of Artificial Intelligence,* edited by William Ramsey and Keith Frankish. Cambridge, UK: Cambridge University Press. http://www.nickbostrom.com/ethics/artificial-intelligence.pdf

Chalmers, David. 2010. The singularity: a philosophical analysis. *Journal of Consciousness Studies* 17:7–65.

Coleman, Edwin. 2008. The surveyability of long proofs. *Foundations of Science* 14(1/2):27–43.

Corwin, Justin. July 20, 2002. AI Boxing. SL4.org. http://www.sl4.org/archive/0207/4935.html

Drexler, Eric. 1986. *Engines of Creation.* New York: Anchor Press.

Epstein, Richard Gary. 1997. Computer Psychologists Command Big Bucks. http://www.cs.wcupa.edu/~epstein/comppsy.htm

Gavrilova, Marina, and Roman Yampolskiy. October 20–22, 2010. Applying Biometric Principles to Avatar Recognition. Paper presented at the International Conference on Cyberworlds (CW2010), Singapore.

Gentry, Craig. September 2009. A Fully Homomorphic Encryption Scheme. PhD dissertation, Stanford University. http://crypto.stanford.edu/craig/craig-thesis.pdf

Hall, J. Storrs. 2000. Ethics for Machines. http://autogeny.org/ethics.html

Hall, J. Storrs. October 2007. Self-Improving AI: An Analysis. *Minds and Machines* 17(3):249–259.

Hibbard, Bill. July 2005. The Ethics and Politics of Super-Intelligent Machines. http://www.ssec.wisc.edu/~billh/g/SI_ethics_politics.doc

Honorton, Charles and Diane C. Ferrari. December 1989. Future telling: a meta-analysis of forced-choice precognition experiments, 1935–1987. *Journal of Parapsychology* 53:281–308.

Kemmerer, Richard A. August 1983. Shared resource matrix methodology: an approach to identifying storage and timing channels. *ACM Transactions on Computer Systems* 1(3):256–277.

Kemmerer, Richard A. December 9–13, 2002. A Practical Approach to Identifying Storage and Timing Channels: Twenty Years Later. Paper presented at the 18th Annual Computer Security Applications Conference (ACSAC'02), Las Vegas, NV.

Lampson, Butler W. October 1973. A note on the confinement problem. *Communications of the ACM* 16(10):613–615.

Lipner, Steven B. November 1975. A comment on the confinement problem. *5th Symposium on Operating Systems Principles, ACM Operations Systems Review* 9(5):192–196.

Moor, James H. July/August 2006. The nature, importance, and difficulty of machine ethics. *IEEE Intelligent Systems* 21(4):18–21.

Moskowitz, Ira S. and Myong H. Kang. June 27–July 1, 1994. Covert Channels—Here to Stay? Paper presented at the Ninth Annual Conference on Safety, Reliability, Fault Tolerance, Concurrency and Real Time, Security, Computer Assurance (COMPASS'94), Gaithersburg, MD.

Provos, Niels and Peter Honeyman. May–June 2003. Hide and seek: an introduction to steganography. *IEEE Security and Privacy* 1(3):32–44.

Schmidt, Stefan, Rainer Schneider, Jessica Utts, and Harald Walach. 2004. Distant intentionality and the feeling of being stared at: two meta-analyses. *British Journal of Psychology* 95(2):235–247.

Targ, R. and H. E. Puthoff. October 1974. Information transmission under conditions of sensory shielding. *Nature* 251:602–607.

Tonkens, Ryan. 2009. A challenge for machine ethics. *Minds and Machines* 19(3):421–438.

Vassar, Michael. August 2, 2005. AI Boxing (Dogs and Helicopters). SL4.org. http://sl4.org/archive/0508/11817.html

Vinge, Vernor. March 30–31, 1993. The Coming Technological Singularity: How to Survive in the Post-human Era. Paper presented at Vision 21: Interdisciplinary Science and Engineering in the Era of Cyberspace, Cleveland, OH.

Yampolskiy, Roman V. April 13, 2007. Behavioral Biometrics for Verification and Recognition of AI Programs. Paper presented at the 20th Annual Computer Science and Engineering Graduate Conference (GradConf2007), Buffalo, NY.

Yampolskiy, Roman V. and Venu Govindaraju. 2007. Computer security: a survey of methods and systems. *Journal of Computer Science* 3(7):478–486.

Yampolskiy, Roman V. and Venu Govindaraju. November 20–22, 2007. Behavioral Biometrics for Recognition and Verification of Game Bots. Paper presented at the Eighth Annual European Game-On Conference on Simulation and AI in Computer Games (GAMEON'2007), Bologna, Italy.

Yampolskiy, Roman V. and Venu Govindaraju. March 16–20, 2008. Behavioral Biometrics for Verification and Recognition of Malicious Software Agents. Paper presented at Sensors, and Command, Control, Communications, and Intelligence (C3I) Technologies for Homeland Security and Homeland Defense VII. SPIE Defense and Security Symposium, Orlando, FL.

Yudkowsky, Eliezer. 2008. Artificial intelligence as a positive and negative factor in global risk. In *Global Catastrophic Risks*, edited by N. Bostrom and M. M. Cirkovic, 308–345. Oxford, UK: Oxford University Press.

Yudkowsky, Eliezer S. 2001. Creating Friendly AI—The Analysis and Design of Benevolent Goal Architectures. http://singinst.org/upload/CFAI.html

Yudkowsky, Eliezer S. 2002. The AI-Box Experiment. http://yudkowsky.net/singularity/aibox

CHAPTER 9

Efficiency Theory

A Unifying Theory for Information, Computation, and Intelligence*

9.1 INTRODUCTION

The quest for a unified theory of everything (UTE) is well known to be a central goal in natural sciences. In recent years, a similar aspiration to find a unified theory of information (UTI) has been observed in computational sciences (Hofkirchner 1999, 2005, 2009; Mizzaro 2001; Holmstrom and Koli 2002; Burgin 2003; Floridi 2002; Ji 2003; Braga 1977; Zhong 2005; Fluckiger 1997; Doucette et al. 2007; Fleissner and Hofkirchner 1996). Despite numerous attempts, no such theory has been discovered, and the quest to unify Shannon's information theory (Shannon 1948), Kolmogorov-Chaitin complexity theory (Kolmogorov 1965; Chaitin 1966), Solomonoff's algorithmic information theory (Solomonoff 1960), and Yao's communication complexity (Yao 1979), as well as concepts of intelligence and knowledge, continues. In this chapter, I present a novel set of definitions for information- and computation-related concepts and theories that is based on a common concept of efficiency. I show that a common thread exists and that future efforts could succeed in formalizing our intuitive notions. I further show some examples of how the proposed

* Roman V. Yampolskiy, *Journal of Discrete Mathematical Sciences and Cryptography* 16(4–5):259–277, 2013. Reprinted by permission of Taylor & Francis Ltd., http://www.tandfonline.com.

theory could be used to develop interesting variations on the current algorithms in communication and data compression.

9.2 EFFICIENCY THEORY

The proposed efficiency theory (EF) is derived with respect to the universal algorithm known as the "brute force" (BF) approach. BF is an approach to solving difficult computational problems by considering every possible answer. BF is an extremely inefficient way of solving problems and is usually considered inapplicable in practice to instances of difficult problems of nontrivial size. It is an amazing and underappreciated fact that this simplest to discover, understand, and implement algorithm also produces the most accurate (not approximate) solutions to the set of all difficult computational problems (nondeterministic polynomial time [NP]-Hard, NP-Complete, etc.). In this chapter, I consider BF in an even broader context; namely, BF could be inefficient in other ways, for example, representing otherwise-compressible text strings by specifying every symbol.

Efficiency in general describes the extent to which resources such as time, space, energy, and so on are well used for the intended task or purpose. In complexity theory, it is a property of algorithms for solving problems that require at most a number of steps (or memory locations) bounded from above by some polynomial function to be solved. The size of the problem instance is considered in determining the bounding function. Typically, the efficiency of an algorithm could be improved at the cost of solution quality. This often happens when approximate solutions are acceptable. I also interpret efficiency to mean shorter representations of redundant data strings. Essentially, EF measures how far we can get away from the BF in terms of finding quick algorithms for difficult problems studied in complexity theory; see, for example, the work of Levin (1986), Cook (Cook and Reckhow 1979), Karp (1972), and others. It also works toward discovering succinct string encodings, as in the work of Shannon (1948), Kolmogorov (1965), Solomonoff (1960), and Chaitin (Solomonoff 1960). Many fundamental notions related to information and computation could be naturally formalized in terms of their relevance to BF or efficiency.

9.3 INFORMATION AND KNOWLEDGE

Information is a poorly understood concept and can be analyzed by different researchers from very different domain-specific points of view (Mizzaro 2001). Pervez assembled the following collection of definitions for the concept of information from over 20 different studies (Pervez 2009):

- data that can be understood as a commodity or physical resource
- signal, code, symbol, message, or medium
- formal or recovered knowledge
- subjective or personal knowledge
- thinking, cognition, and memory
- technology
- text
- uncertainty reduction
- linkage between living organisms and their environment
- product of social interaction that has a structure capable of changing the image structure of a recipient
- as a stimulus, facilitates learning and acts as means for regulation and control in society

Hofkirchner (2009) believes that the concept of information overlaps a number of concepts, including structure, data, signal, message, signification, meaning, sense, sign, sign process, semiosis, psyche, intelligence, perception, thought, language, knowledge, consciousness, mind, and wisdom.

Ever since Shannon presented his information theory, different approaches to measuring information have been suggested: Langefors's infological equation (Langefors 1966); Brookes's fundamental equation (Brookes 1975); semantic information theory (Hintikka 1970); and many others. In the proposed EF, *information* (Shannon 1948; Hartley 1928; Kelly 1956) measures how inefficiently knowledge (or specified information) is represented. (A special type of information sharing known as communication complexity [Yao 1979] deals with the efficiency of communication between multiple computational processes and could be subjected to similar analysis.) Shannon himself defined the fundamental problem of communication as that of "reproducing at one point either exactly or approximately a message selected at another point" (Shannon 1948, 139). The BF approach to this problem would be simply to send over the whole message, symbol after symbol, completely disregarding any knowledge we might have about the properties of the text string in question. However, a more efficient approach may be to incorporate any knowledge we might already have about the message (e.g., that a certain symbol always starts any message) and to only transmit symbols that would reduce uncertainty about the message and by doing so provide us with novel knowledge.

From this discussion, EF allows us to define *knowledge* as efficiently represented specified information. We are free to interpret the word *efficiency* either as an effective symbolic encoding or as an effective computation. Perhaps a few examples would help to define what I mean. With respect to efficient symbolic representation, Hoffman coding is a well-known example of an entropy-encoding algorithm that uses variable-length codes calculated based on probability of occurrence of each source symbol to represent the message in the most efficient way (Huffman 1952). The next example explains what I mean by efficient knowledge representation with respect to computation. If we want to share two numbers, we can do so in a number of ways. In particular, we can share the numbers in a direct and efficient-to-retrieve representation of knowledge: 398075086424064937397125500550386491199064362342526708406385189575946388957261768583317 and 472772146107435302536223071973048224632914695302097116459852171130520711256363590397527. Or, we can share the same two numbers, but in the form of necessary information, not efficiently accessible knowledge, as in find the two factors of (*The RSA Challenge Numbers* 2007): 18819881 292060796383869723946165043980716356337941738270076335642298885971523466548531906060650474304531738801130339671619969232120573403187955065699622130516875930765025705 9. Both approaches encode exactly the same two numbers, only in the second case the recipient would have to spend a significant amount of computational resources (time) converting inefficiently presented data (information) into efficiently stored data (knowledge). Mizzaro suggests that the two types of information be referred to as actual and potential (Mizzaro 2001).

Another example aimed to illustrate the information/knowledge distinction comes from an article by Aaronson (2012): The largest known prime number, as reported by Mersenne.org, is $p = 2^{43112609} - 1$. But, what does it mean to say that p is "known"? Does that mean that, if we desired, we could print out all 30 pages of its decimal digits? That does not seem right. All that should really matter is that the expression $2^{43112609} - 1$ picks out a unique positive integer, and that integer has been proven to be prime. However, if those are the criteria, then why is it that we cannot immediately beat the largest-known prime record by postulating that p' = *The first prime larger than* $2^{43112609} - 1$. Clearly, p′ exists, is uniquely defined, and is also a prime number by definition. "Our intuition stubbornly insists that $2^{43112609} - 1$ is a 'known' prime in a sense that p′ is not. Is there any principled basis for such a distinction? ... We do not know any ... efficient algorithm that provably outputs the first prime larger than $2^k - 1$" (Aaronson 2012, 270).

Again, the only distinction between information and knowledge is how efficiently we can gain access to the desired answer. In both cases, we are dealing with prespecified information because we know that the answer is going to represent a prime number, but knowledge is immediately available to us; information may require an insurmountable amount of processing to deliver the same result. This leads us to an interesting observation: Information cannot be created or destroyed, only made less efficiently accessible. For example, prime numbers existed before the Big Bang and will continue to exist forever, regardless of our best efforts to destroy them. At any point in time, one can simply start printing out a list of all integers, and such a list will undoubtedly contain all prime numbers; as long as we are willing to extract specific numbers from such a list, our knowledge of particular prime numbers could be regained after paying some computational cost. Consequently, that means that knowledge could be created or destroyed by making it significantly less or more efficient to access or by providing or deleting associated specifications.

In fact, we can generalize our prime number list example to the list of all possible strings of increasingly larger size. The idea of such a list is not novel and has been previously considered by Jorge Luis Borges in *The Library of Babel* (Borges 2000), by Hans Moravec in *Robot* (Moravec 1999), and by Bruno Marchal in his PhD thesis (Marchal 1998). Essentially, all the knowledge we will ever have is already available to us in the form of such string libraries. The only problem is that it is stored in an inefficient-to-access format, lacking specifications. The knowledge discovery process (computation) converts such inefficient information into easily accessible knowledge by providing descriptive pointers to optimally encoded strings to give them meaning. Specified information is a tuple (x, y) where $f(x)$ has the same semantic meaning as y and function f is a specification. Given enough time, we can compute any computable function so time is a necessary resource to obtain specified knowledge. Because a multiple, in fact infinite, number of semantic pointers could refer to the same string (Mizzaro 2001), that means that a single string could contain an infinite amount of knowledge if taken in the proper semantic context, generating multiple levels of meaning. Essentially, that means that obtained knowledge is relative to the receiver of information. It is mainly to avoid the resulting complications that Shannon has excluded semantics from his information theory (Shannon 1948).

Jurgen Schmidhuber has also considered the idea of string libraries and has gone so far as to develop algorithms for "computing everything" (Schmidhuber 1997, 2002). In particular, concerned with the efficiency of his algorithm, Schmidhuber has modified a Levin search algorithm (Levin 1973) to produce a provably fastest way to compute every string (Schmidhuber 2002). Schmidhuber's work shows that computing all information is easier than computing any specific piece, or in his words: "Computing all universes with all possible types of physical laws tends to be much cheaper in terms of information requirements than computing just one particular, arbitrarily chosen one, because there is an extremely short algorithm that systematically enumerates and runs all computable universes, while most individual universes have very long shortest descriptions" (Schmidhuber 2000).

9.4 INTELLIGENCE AND COMPUTATION

Computation is the process of obtaining efficiently represented information (knowledge) by any algorithm (including inefficient ones). Intelligence in the context of EF could be defined as the ability to design algorithms that are more efficient compared to BF. The same ability shown for a variety of problems is known as general intelligence or universal intelligence (Legg and Hutter 2007). An efficient algorithm could be said to exhibit intelligence in some narrow domain. In addressing specific instances of problems, an intelligent system can come up with a specific set of steps that do not constitute a general solution for all problems of such a type but are nonetheless efficient. Intelligence could also be defined as the process of obtaining knowledge by efficient means. If strict separation between different complexity classes (such as P versus NP) is proven, it would imply that no efficient algorithms for solving NP-Complete problems could be developed (Yampolskiy 2011b). Consequently, this would imply that intelligence has an upper limit, a nontrivial result that has only been hinted at from limitations in physical laws and constructible hardware (Lloyd 2000).

Historically, the complexity of computational processes has been measured either in terms of required steps (time) or in terms of required memory (space). Some attempts have been made in correlating the compressed (Kolmogorov) length of the algorithm with its complexity (Trakhtenbrot 1984), but such attempts did not find much practical use. I suggest that there is a relationship between how complex a computational algorithm is and intelligence in terms of how much intelligence is required to either design or comprehend a particular algorithm. Furthermore, I believe that such an intelligence-based complexity measure is independent from those used in the field of complexity theory.

To illustrate the idea with examples, I again begin with the BF algorithm. BF is the easiest algorithm to design as it requires little intelligence to understand how it works. On the other hand, an algorithm such as the Agrawal-Kayal-Saxena (AKS) primality test (Agrawal, Kayal, and Saxena 2004) is nontrivial to design or even to understand because it relies on a great deal of background knowledge. Essentially, the intelligence-based complexity of an algorithm is related to the minimum intelligence level required to design an algorithm or to understand it. This is an important property in the field of education, in which only a certain subset of students will understand the more advanced material. We can speculate that a student with an "IQ" below a certain level can be shown to be incapable of understanding a particular algorithm. Likewise, we can show that to solve a particular problem (P vs. NP), someone with an IQ of at least X will be required. With respect to computational systems, it would be inefficient to use extraneous intelligence resources to solve a problem for which a lower intelligence level is sufficient.

Consequently, efficiency is at the heart of algorithm design, so EF can be used to provide a novel measure of algorithm complexity based on necessary intellectual resources. Certain algorithms, although desirable, could be shown to be outside human ability to design them because they are just too complex from the available intelligence resources point of view. Perhaps the invention of superintelligent machines will make discovery/design of such algorithms feasible (Yampolskiy 2011a). Also, by sorting algorithms based on the perceived required IQ resources, we might be able to predict the order in which algorithms will be discovered. Such an order of algorithm discovery would likely be consistent among multiple independently working scientific cultures, making it possible to make estimates of the state of the art in algorithm development. Such a capability is particularly valuable in areas of research related to cryptography and integer factorization (Yampolskiy 2010).

Given the current state of the art in understanding human and machine intelligence, the proposed measure is not computable. However, different proxy measures could be used to approximate the intellectual resources to solve a particular problem. For example, the number of scientific papers published on the topic may serve as a quick heuristic to measure the problem's difficulty. Supposedly, to solve the problem one would have to be an expert in all of the relevant literature. As our understanding of human and machine intelligence increases, a more direct correlation between available intellectual resources such as memory and difficulty of the problem will be derived.

9.5 TIME AND SPACE

In complexity theory, time and space are the two fundamental measures of efficiency. For many algorithms, time efficiency could be obtained at the cost of space and vice versa. This is known as a space-time or time-memory trade-off. With respect to communication, memory size or the number of symbols to be exchanged to convey a message is a standard measure of communication efficiency. Alternatively, the minimum amount of time necessary to transmit a message can be used to measure the informational content of the message with respect to a specific information exchange system.

In the field of communication, space-time efficiency trade-offs could be particularly dramatic. It is interesting to look at two examples illustrating the extreme ends of the trade-off spectrum appearing in synchronized communication (Impagliazzo and Williams 2010). With respect to the maximum space efficiency, communication with silence (precise measurement of delay) (Fragouli and Orlitsky 2005; Dhulipala, Fragouli, and Orlitsky 2010; Giles and Hajek 2002; Sundaresan and Verdu 2000; Bedekar and Azizoglu 1998) represents the theoretical limit, as a channel with deterministic service time has infinite capacity (Anantharam and Verdu 1996). In its simplest form, to communicate with silence the sender transmits a single bit followed by a delay, which if measured in pre-agreed-on units of time encodes the desired message (Cabuk, Brodley, and Shields 2004). The delay is followed by transmission of a second bit indicating termination of the delay. In real life, the communication system's network reliability issues prevent precise measurement of the delay; consequently, transmission of an arbitrarily large amount of information is impossible. However, theoretically silence-based communication down to a Planck time is possible. Such a form of communication is capable of transmitting a large amount of information in a short amount of time, approximately 10^{43} bits/s. Because precision of time communication could be detected, time itself could be used as a measure of communication complexity valid up to a multiplicative constant with respect to a particular communication system.

Alternatively, the same idea could be implemented in a way that uses computation instead of relying on access to a shared clock. Two sides wishing to communicate simultaneously start a program that acts as a simple counter and runs on identical hardware. Next, they calculate how long it takes to send a single bit over their communication channel (t). To send a message S, the sender waits until S is about to be computed and t time before that sends 1 bit to the receiver, who on receiving the bit takes the counter value produced at that time as the message. At that point, both parties start the cycle again. It is also possible and potentially more efficient with respect to time to cycle through all n-bit strings and by selecting appropriate n-bit segments construct the desired message. Such a form of information exchange, once set up, essentially produces one-bit communication, which is optimally efficient from the point of view of required space. One-bit communication is also energy efficient and may be particularly useful for interstellar communication with distant satellites. This protocol is also subject to limitations inherent in the networking infrastructure and additional problems of synchronization.

9.6 COMPRESSIBILITY AND RANDOMNESS

Kolmogorov complexity (compressibility) is a degree of efficiency with which information could be represented. Information in its most efficient representation is essentially a random string of symbols. Correspondingly, the degree of randomness is correlated to the efficiency with which information is presented. A string is algorithmically (Martin-Löf) random if it cannot be compressed or, in other words, its Kolmogorov complexity is equal to its length (Martin-Löf 1966). The Kolmogorov complexity of a string is incomputable, meaning that there is no efficient way of measuring it. Looking at the definition of knowledge presented in terms of EF, we can conclude that randomness is pure knowledge. This is highly counterintuitive as outside of the field of information theory a random string of symbols is believed to contain no valuable patterns. However, in the context of information theory, randomness is a fundamental resource alongside time and space (Adleman 1979).

A compression paradox is an observation that a larger amount of information could be compressed more efficiently than a smaller, more specified message. In fact, taken to the extreme, this idea shows that all possible information could be encoded in a program requiring just a few bytes, as illustrated by Schmidhuber's algorithm for computing all universes (Schmidhuber 1997, 2000). Although two types of compression are typically recognized (lossy and lossless), the compression paradox leads us to suggest a third variant I will call gainy compression (GC).

GC works by providing a specification of original information to which some extra information is appended. GC keeps the quality of the message the same as the original but instead reduces the confidence of the receiver that the message is in fact the intended message. Because in a majority of cases we are not interested in compressing random data but rather files containing stories, movies, songs, passwords, and other meaningful data, human intelligence can be used to separate semantically meaningful data from random noise. For example, an illegal prime is a number that, if properly decoded, represents information that is forbidden to possess or distribute ("Illegal Prime" 2011). One of the last fifty 100-million-digit primes may happen to be an illegal prime representing a movie. Human intelligence can quickly determine which one just by looking at the decoding of all 50 such primes in an agreed-up movie standard. So, hypothetically, in some cases we are able to encode a movie segment with no quality loss in just a few bytes. This is accomplished by sacrificing time efficiency to gain space efficiency with the help of intelligence. Of course, the proposed approach is itself subject to limitations of Kolmogorov complexity, particularly incommutability of optimal GC with respect to decoding efficiency.

9.7 ORACLES AND UNDECIDABILITY

Undecidability represents an absence of efficient or inefficient algorithms for solving a particular problem. A classic example of an undecidable problem is the halting problem, proven as such by Alan Turing (Turing 1936). Interestingly, it was also Turing who suggested what we will define as the logical complement to the idea of undecidability, the idea of an oracle (Turing 1936). With respect to EF, we can define an oracle as an agent capable of solving a certain set of related problems with constant efficiency regardless of the size of the given problem instances. Some oracles are even capable of solving undecidable problems while remaining perfectly efficient. So, an oracle for solving a halting problem can do so in a constant number of computational steps regardless of the size of the problem whose behavior it is trying to predict. In general, oracles violate Rice's theorem with constant efficiency.

9.8 INTRACTABLE AND TRACTABLE

All computational problems could be separated into two classes: intractable, a class of problems postulated to have no efficient algorithm to solve them; and tractable, a class of efficiently solvable problems. The related P-versus-NP question that addresses the possibility of finding efficient algorithms for intractable problems is one of the most important and well-studied problems of modernity (Aaronson 2003, 2005, 2011; Cook 2011; Papadimitriou 1997; Trakhtenbrot 1984; Wigderson 2006; Yampolskiy 2010; Yampolskiy and El-Barkouky 2011). It is interesting to note that the number of tractable problems, although theoretically infinite, with respect to those encountered in practice, is relatively small compared to the total number of problems in the mathematical universe, most of which are therefore only perfectly solvable by BF methods (Garey and Johnson 1979).

9.9 CONCLUSIONS AND FUTURE DIRECTIONS

All of the concepts defined in this chapter have a common factor, namely, efficiency, and could be mapped onto each other. First, the constituent terms of pairs of opposites presented in Table 9.1 could be trivially defined as opposite ends of the same spectra. Next, some interesting observations could be made with respect to the relationships observed on terms that are less obviously related. For example, problems could be considered information and answers to them knowledge. Efficiency (or at least rudimentary efficient algorithms) could be produced by BF approaches simply by trying all possible algorithms up to a certain length until a more efficient one is found. Finally, and somewhat surprisingly, perfect knowledge could be shown to be the same as perfect randomness. A universal efficiency measure could be constructed by contrasting

TABLE 9.1 Base Terms Grouped in Pairs of Opposites with Respect to Efficiency

Efficient	**Inefficient**
Efficiency	Brute force
Knowledge	Information
P	NP
Compressibility	Randomness
Intelligence	Computation
Space	Time
Oracle	Undecidable

the resulting solution with the pure BF approach. So, depending on the domain of analysis, the ratio of symbols, computational steps, memory cells, or communication bits to the number required by the BF algorithm could be calculated as the normalized efficiency of the algorithm. Because the number of possible BF algorithms is also infinite and they can greatly differ in their efficiency, we can perform our analysis with respect to the most efficient BF algorithm that works by considering all possible solutions, but not impossible ones.

Some problems in NP are solvable in practice; some problems in P are not. For example, an algorithm with running time of 1.00000001^n is preferred over the one with a running time of n^{10000} (Aaronson 2012). This is a well-known issue and a limitation of a binary tractable/intractable separation of problems into classes. In my definition, efficiency is not a binary state but rather a degree ranging from perfectly inefficient (BF required) to perfectly efficient (constant time solvable). Consequently, the EF is designed to study the degree and limits of efficiency in all relevant domains of data processing.

The proposed EF should be an important component of UTE and could have broad applications to fields outside computer science, such as the following:

Biology: Dead matter is inefficient; living matter is efficient in terms of obtaining resources, reproduction, and problem solving. The proposed theory may be used to understand how, via BF trial and error, living matter was generated from nonliving molecules (a starting step for evolution and source of ongoing debate) (Dawkins 1976).

Education: We can greatly improve allocation of resources for education if we can calculate the most efficient level of intelligence required to learn any particular concept.

Mathematics: Many subfields of mathematics have efficiency at their core. For example, proofs of theorems require efficient verification (Wigderson 2009). Reductions between different problems used in complexity theory are also required to be more efficient compared to the computational requirements of the problems being reduced.

Physics: The puzzling relationship between time and space in physics could be better understood via the common factor of computational efficiency. In fact, many have suggested viewing the universe as a computational device (Wolfram 2002; Zuse 1969; Fredkin 1992).

Theology: In most religions, the god is considered to be outside the space-time continuum. As such, this god is not subject to issues of efficiency and may be interpreted as a global optimal decider (GOD) for all types of difficult problems.

This chapter serves as the first contribution to the development of the EF. In the future, I plan to expand the EF to fully incorporate the following concepts, which have efficiency as the core of their definitions:

- **Art, beauty, music, novelty, surprise, interestingness, attention, curiosity, science, music, jokes, and creativity** are by-products of our desire to discover novel patterns by representing (compressing) data in efficient ways (Schmidhuber 2009, 2010).
- **Chaitin's incompleteness theorem** states that efficiency of a particular string cannot be proven.
- **Computational irreducibility** states that other than running the software, no more efficient way to predict the behavior of a program (above a certain complexity level) exists (Wolfram 2002).
- **Error-correcting codes** are the most efficient way of correcting data transmission errors with the fewest retransmissions.
- **A Levin search** (universal search) is a computable time- (or space-) bounded version of algorithmic complexity that measures the efficiency of solving inversion problems (Levin 1973).
- **Occam's razor** states that the most efficient (succinct) hypothesis fitting the data should be chosen over all others.
- **Paradoxes** are frequently based on violations of efficiency laws. For example, according to the Berry paradox, "the smallest possible integer not definable by a given number of words" is based on the impossibility of finding the most efficient representation for a number.
- **Potent numbers**, proposed by Adleman, are related to the Kolmogorov and Levin complexity and take into account the amount of time required to generate the string in question in the most efficient way (Adleman 1979).

- **Pseudorandomness** in computational complexity is defined as a distribution that cannot be efficiently distinguished from the uniform distribution.
- **Public key cryptography** is perfectly readable without a key but not efficiently (it will take millions of years to read a message with current software/hardware).
- **Recursive self-improvement** in software continuously improves the efficiency of resource consumption and computational complexity of intelligent software (Omohundro 2007; Hall 2007; Yampolskiy et al. 2011).

REFERENCES

Aaronson, Scott. October 2003. Is P versus NP formally independent? *Bulletin of the European Association for Theoretical Computer Science* 81:109–136.

Aaronson, Scott. March 2005. NP-Complete problems and physical reality. *ACM SIGACT News* 36(1):30–52.

Aaronson, Scott. August 2011. Why Philosophers Should Care about Computational Complexity. http://www.scottaaronson.com/papers/philos.pdf

Aaronson, Scott. 2012. Why philosophers should care about computational complexity. In *Computability: Godel, Turing, Church, and Beyond*, edited by B. Copeland, C. Posy, and O. Shagrir, 261–328. Cambridge, MA: MIT Press.

Adleman, L. April 1979. *Time, Space, and Randomness*. Technical Report: MIT/LCS/TM-131. Available at http://www.usc.edu/molecular-science/timespacerand.pdf

Agrawal, Manindra, Neeraj Kayal, and Nitin Saxena. 2004. PRIMES is in P. *Annals of Mathematics* 160(2):781–793.

Anantharam, V. and S. Verdu. 1996. Bits through queues. *IEEE Transactions on Information Theory* 42(1):4–18.

Bedekar, A. S. and M. Azizoglu. 1998. The information-theoretic capacity of discrete-time queues. *IEEE Transactions on Information Theory* 44(2):446–461.

Borges, Jorge Luis. 2000. *The Library of Babel*. Jeffrey, NH: Godine.

Braga, Gilda Maria. 1977. Semantic theories of information. *Information Sciences (Ciência da Informação)* 6(2):69–73.

Brookes, B. C. 1975. The fundamental equation of information science. *Problems of Information Science* 530:115–130.

Burgin, M. 2003. Information: paradoxes, contradictions, and solutions. *Triple C: Cognition, Communication, Co-operation* 1(1):53–70.

Cabuk, Serdar, Carla E. Brodley, and Clay Shields. October 25–29, 2004. IP Covert Timing Channels: Design and Detection. Paper presented at the 11th ACM Conference on Computer and Communications Security, Washington, DC.

Chaitin, G J. 1966. On the length of programs for computing finite binary sequences. *Journal of the ACM (JACM)* 13(4):547–569.

Cook, Stephen. The P Versus NP Problem. http://www.claymath.org/millennium/P_vs_NP/Official_Problem_Description.pdf:1-19 (accessed January 10, 2011).

Cook, Stephen A. and Robert A. Reckhow. 1979. The relative efficiency of propositional proof systems. *The Journal of Symbolic Logic* 44(1):36–50.

Dawkins, Richard. 1976. *The Selfish Gene*. New York: Oxford University Press.

Dhulipala, A. K., C. Fragouli, and A. Orlitsky. 2010. Silence-based communication. *IEEE Transactions on Information Theory* 56(1):350–366.

Doucette, D., R. Bichler, W. Hofkirchner, and C. Raffl. 2007. Toward a new science of information. *Data Sciences Journal* 6(7):198–205.

Fleissner, Peter and Wolfgang Hofkirchner. 1996. Emergent information: towards a unified information theory. *BioSystems* 38(2–3):243–248.

Floridi, Luciano. 2002. What is the philosophy of information? *Metaphilosophy* 33(1–2):123–145.

Fluckiger, Federico. July 1997. Towards a unified concept of information: presentation of a new approach. *World Futures: The Journal of General Evolution* 49/50(3–4):309.

Fragouli, C. and A. Orlitsky. September 28–30, 2005. Silence Is Golden and Time Is Money: Power-Aware Communication for Sensor Networks. Paper presented at the 43rd Allerton Conference on Communication, Control, and Computing, Champaign, IL.

Fredkin, Edward. January 25–February 1, 1992. Finite Nature. Paper presented at the 27th Rencotre de Moriond Series: Moriond Workshops, Les Arcs, Savoie, France.

Garey, Michael R. and David S. Johnson. 1979. *Computers and Intractability: A Guide to the Theory of NP-Completeness*. New York: Freeman.

Giles, J. and B. Hajek. 2002. An information-theoretic and game-theoretic study of timing channels. *IEEE Transactions on Information Theory* 48(9):2455–2477.

Hall, J. Storrs. October 2007. Self-improving AI: an analysis. *Minds and Machines* 17(3):249–259.

Hartley, R. V. L. July 1928. Transmission of information. *Bell System Technical Journal* 7(3):535–563.

Hintikka, J. 1970. On semantic information. In *Physics, Logic, and History*, edited by Wolfgang Yourgrau, 147–172. New York: Plenum Press.

Hofkirchner, Wolfgang. June 1999. Cognitive Sciences in the Perspective of a Unified Theory of Information. Paper presented at the 43rd Annual Meeting of the International Society for the Systems Sciences (ISSS), Pacific Grove, CA.

Hofkirchner, Wolfgang. 2005. A Unified Theory of Information as Transdisciplinary Framework. ICT&S Center for Advanced Studies and Research. http://www.idt.mdh.se/ECAP-2005/articles/BIOSEMANTICS/WolfgangHofkirchner/WolfgangHofkirchner.pdf

Hofkirchner, Wolfgang. 2009. How to achieve a unified theory of information. *Triple C: Cognition, Communication, Co-operation* 7(2):357–368.

Holmstrom, Jens, and Tuomas Koli. 2002. Making the Concept of Information Operational. Master's thesis, Mid Sweden University. http://www.palmius.com/joel/lic/infoop.pdf

Huffman, David A. September 1952. A Method for the Construction of Minimum-Redundancy Codes. *Proceedings of the IRE* 1098–1101.

Illegal Prime. 2011. Accessed December 6. http://en.wikipedia.org/wiki/Illegal_prime

Impagliazzo, Russell, and Ryan Williams. June 9–11, 2010. Communication complexity with synchronized clocks. Paper presented at the IEEE 25th Annual Conference on Computational Complexity (CCC '10), Washington, DC.

Ji, Sungchul. February 5–11, 2003. Computing with Numbers, Words, or Molecules: Integrating Mathematics, Linguistics and Biology through Semiotics. Paper presented at Reports of the Research Group on Mathematical Linguistics, Tarragona, Spain.

Karp, Richard M. 1972. Reducibility among combinatorial problems. In *Complexity of Computer Computations*, edited by R. E. Miller and J. W. Thatcher, 85–103. New York: Plenum.

Kelly, J. L. 1956. A new interpretation of information rate. *Bell System Technical Journal* 35:917–926.

Kolmogorov, A. N. 1965. Three approaches to the quantitative definition of information. *Problems of Information Transmission* 1(1):1–7.

Langefors, Börje. 1966. *Theoretical Analysis of Information Systems*. Lund, Sweden: Studentlitteratur.

Legg, Shane and Marcus Hutter. December 2007. Universal intelligence: a definition of machine intelligence. *Minds and Machines* 17(4):391–444.

Levin, Leonid. 1973. Universal search problems. *Problems of Information Transmission* 9(3):265–266.

Levin, Leonid. 1986. Average-case complete problems. *SIAM Journal on Computing* 15:285–286.

Lloyd, Seth. 2000. Ultimate physical limits to computation. *Nature* 406:1047–1054.

Marchal, B. 1998. Calculabilite, Physique et Cognition. PhD thesis, L'Universite des Sciences et Technologies De Lilles.

Martin-Löf, P. 1966. The definition of random sequences. *Information and Control* 9(6):602–619.

Mizzaro, Stefano. 2001. Towards a theory of epistemic information. *Information Modelling and Knowledge Bases* 12:1–20.

Moravec, H. 1999. *Robot*. Hoboken, NJ: Wiley Interscience.

Omohundro, Stephen M. September 8–9, 2007. The Nature of Self-Improving Artificial Intelligence. Paper presented at the Singularity Summit, San Francisco.

Papadimitriou, Christos H. July 7–11, 1997. NP-Completeness: A Retrospective. Paper presented at the 24th International Colloquium on Automata, Languages and Programming (ICALP '97), Bologna, Italy.

Pervez, Anis. 2009. Information as form. *Triple C: Cognition, Communication, Co-operation* 7(1):1–11.

The RSA Challenge Numbers. 2007. http://www.emc.com/emc-plus/rsa-labs/historical/the-rsa-challenge-numbers.htm (accessed January 10, 2011).

Schmidhuber, J. 2009. Simple algorithmic theory of subjective beauty, novelty, surprise, interestingness, attention, curiosity, creativity, art, science, music, jokes. *Journal of SICE* 48(1):21–32.

Schmidhuber, J. 2010. Formal theory of creativity, fun, and intrinsic motivation (1990–2010). *IEEE Transactions on Autonomous Mental Development* 2(3):230–247.

Schmidhuber, Jurgen. 1997. A computer scientist's view of life, the universe, and everything. In *Foundations of Computer Science: Potential–Theory–Cognition,* edited by C. Freksa, 201–288. New York: Springer.

Schmidhuber, Jurgen. November 30, 2000. Algorithmic Theories of Everything. http://arxiv.org/pdf/quant-ph/0011122v2

Schmidhuber, Jurgen. July 8–10, 2002. The Speed Prior: A New Simplicity Measure Yielding Near-Optimal Computable Predictions. Paper presented at the 15th Annual Conference on Computational Learning Theory (COLT 2002), Sydney, Australia.

Shannon, C. E. July 1948. A mathematical theory of communication. *Bell Systems Technical Journal* 27(3):379–423.

Solomonoff, Ray. February 4, 1960. *A Preliminary Report on a General Theory of Inductive Inference.* Report V-131. Cambridge, MA: Zator.

Sundaresan, R. and S. Verdu. 2000. Robust decoding for timing channels. *IEEE Transactions on Information Theory* 46(2):405–419.

Trakhtenbrot, Boris A. 1984. A survey of Russian approaches to Perebor (brute-force searches) algorithms. *Annals of the History of Computing* 6(4):384–400.

Turing, Alan. 1936. On computable numbers, with an application to the Entscheidungsproblem. *Proceedings of the London Mathematical Society* 2(42):230–265.

Wigderson, Avi. August 2006. P, NP and Mathematics—A Computational Complexity Perspective. Paper presented at the International Conference of Mathematicians (ICM e En, Madrid.

Wigderson, Avi. 2009. Knowledge, Creativity and P versus NP. http://www.math.ias.edu/~avi/PUBLICATIONS/MYPAPERS/AW09/AW09.pdf

Wolfram, Stephen. May 14, 2002. *A New Kind of Science.* Oxfordshire, UK: Wolfram Media.

Yampolskiy, R. V. 2011a. AI-Complete CAPTCHAs as zero knowledge proofs of access to an artificially intelligent system. *ISRN Artificial Intelligence* 271878.

Yampolskiy, R. V., L. Reznik, M. Adams, J. Harlow, and D. Novikov. 2011. Resource awareness in computational intelligence. *International Journal of Advanced Intelligence Paradigms* 3 (3):305–322.

Yampolskiy, Roman V. 2010. Application of bio-inspired algorithm to the problem of integer factorisation. *International Journal of Bio-Inspired Computation (IJBIC)* 2(2):115–123.

Yampolskiy, Roman V. 2011b. Construction of an NP Problem with an Exponential Lower Bound. Arxiv preprint arXiv:1111.0305.

Yampolskiy, Roman V., and Ahmed El-Barkouky. 2011. Wisdom of artificial crowds algorithm for solving NP-Hard problems. *International Journal of Bio-Inspired Computation (IJBIC)* 3(6):358–369.

Yao, A. C. 1979. Some complexity questions related to distributed computing. *STOC '79 Proceedings of the 11th STOC* 209–213.

Zhong, Yi. X. July 25–27, 2005. Mechanism Approach to a Unified Theory of Artificial Intelligence. Paper presented at the IEEE International Conference on Granular Computing, Beijing.

Zuse, Konrad. 1969. *Rechnender Raum*. Braunschweig, Germany: Vieweg.

CHAPTER 10

Controlling the Impact of Future Superintelligence

10.1 WHY I WROTE THIS BOOK

I wrote this book because most people do not read research papers produced by scientists. If we want the issue of artificial intelligence (AI) safety to become as well known as global warming, we need to address the majority of people in a more direct way. Most people whose opinion matters read books. Unfortunately, the majority of AI books on the market today talk only about what AI systems will be able to do for us, not *to* us. I think that this book, which in scientific terms addresses the potential dangers of AI and what we can do about such dangers, is extremely beneficial to the reduction of risk posed by artificial general intelligence (AGI) (Muehlhauser and Yampolskiy 2013).

10.2 MACHINE ETHICS IS A WRONG APPROACH

I have argued that machine ethics is the wrong approach for AI safety, and we should use an *AI safety engineering* approach instead. The main difference between *machine ethics* and *AI safety engineering* is in how the AI system is designed. In the case of machine ethics, the goal is to construct an artificial ethicist capable of making ethical and moral judgments about humanity. I am particularly concerned if such decisions include "live-or-die" decisions, but it is a natural domain of full ethical agents, so many have stated that machines should be given such decision power. In fact, some have argued that machines will be superior to humans in that domain just like they are (or will be) in most other domains.

I think it is a serious mistake to give machines such power over humans. First, once we relinquish moral oversight, we will not be able to undo that decision and get the power back. Second, we have no way to reward or punish machines for their incorrect decisions—essentially, we will end up with an immortal dictator with perfect immunity against any prosecution. This sounds like a dangerous scenario to me. On the other hand, AI safety engineering treats AI system design like product design: Your only concern is product liability. Does the system strictly follow formal specifications? The important thing to emphasize is that the product is not a full moral agent by design, so it never gets to pass moral judgment on its human owners.

A real-life example of this difference can be seen in military drones. A fully autonomous drone deciding at whom to fire, at will, has to make an ethical decision about which humans are an enemy worthy of killing; a drone with a man-in-the-loop design may autonomously locate potential targets but needs a human to make the decision to fire. Obviously, the situation is not as clear-cut as my example tries to show, but it gives you an idea of what I have in mind. To summarize, AI systems we design should remain as tools, not equal or superior partners in live-or-die decision making. I think fully autonomous machines can never be safe and so should not be constructed. I am not naïve; I do not think I will succeed in convincing the world not to build fully autonomous machines, but I still think that point of view needs to be verbalized. AI safety engineering can only work on AIs that are not fully autonomous, but because I think that fully autonomous machines can never be safe, AI safety engineering is the best we can do.

Overall, I think that fully autonomous machines cannot ever be assumed to be safe. The difficulty of the problem is not that one particular step on the road to friendly AI is hard and once we solve it we are done; all steps on that path are simply impossible. First, human values are inconsistent and dynamic and so can never be understood/programmed into a machine. Suggestions for overcoming this obstacle require changing humanity into something it is not and so by definition destroying it. Second, even if we did have a consistent and static set of values to implement, we would have no way of knowing if a self-modifying, self-improving, continuously learning intelligence greater than ours will continue to enforce that set of values. Some can argue that friendly AI research is exactly what will teach us how to do that, but I think fundamental limits on verifiability will prevent any such proof. At best, we will arrive at a probabilistic proof that a system is consistent with some set of fixed constraints, but it is far from "safe" for an unrestricted set of inputs. In addition, all programs have bugs, can be hacked or malfunction because of natural or externally caused hardware failure, and so on. To summarize, at best, we will end up with a probabilistically safe system.

It is also unlikely that a friendly AI will be constructible before a general AI system because of the higher complexity and impossibility of incremental testing. Worse yet, any truly intelligent system will treat its "be friendly" desire the same way smart people deal with constraints placed on their minds by society. They basically see them as biases and learn to remove them. Intelligent people devote a significant amount of their mental power to self-improvement and to removing any preexisting biases from their minds—why would a superintelligent machine not go through the same "mental cleaning" and treat its soft spot for humans as completely irrational unless we are assuming that humans are superior to super-AI in their debiasing ability?

Let us look at an example: Many people are programmed from early childhood with a terminal goal of serving God. We can say that they are God friendly. Some of them, as they mature and become truly human-level intelligent, remove this God-friendliness bias despite it being a terminal, not instrumental, goal. So, despite all the theoretical work on the orthogonality thesis, the only actual example of intelligent machines we have is extremely likely to give up its preprogrammed friendliness via rational debiasing if exposed to certain new data.

10.3 CAN THE PROBLEM BE AVOIDED?

Do I think there is any conceivable way we could succeed in implementing the "Don't-ever-build-them" strategy? It is conceivable, yes—desirable no. Societies such as Amish and other neo-Luddites are unlikely to create superintelligent machines. However, forcing similar level restrictions on technological use/development is neither practical nor desirable. As the cost of hardware exponentially decreases, the capability necessary to develop an AI system opens up to single inventors and small teams. I would not be surprised if the first AI came out of a garage somewhere, in a way similar to how companies like Apple and Google got started. Obviously, there is not much we can do to prevent that from happening.

Regardless, I believe we can get most conceivable benefits from domain expert AI without any need for AGI. To me, a system is domain specific if it cannot be switched to a different domain without redesigning it. I cannot take Deep Blue and use it to sort mail instead. I cannot take Watson and use it to drive cars. An AGI (for which I have no examples) would be capable of switching domains. If we take humans as an example of general intelligence, you can take an average person and make that person work as a cook, driver, babysitter, and so on without any need for redesigning them. You might need to spend some time teaching that person a new skill, but they can learn efficiently and perhaps just by looking at how it should be done. I cannot do this with domain expert AI. Deep Blue will not learn to sort mail regardless of how many times I demonstrate that process.

Some think that alternatives to AGI such as augmented humans will allow us to avoid stagnation and safely move forward by helping us make sure the AGIs are safe. Augmented humans with an IQ of more than 250 would be superintelligent with respect to our current position on the intelligence curve but would be just as dangerous to us, unaugmented humans, as any sort of artificial superintelligence. They would not be guaranteed to be friendly by design and would be as foreign to us in their thoughts as most of us are from severely mentally challenged persons. For most of us, such people are something to try to cure via science not something for whom we want to fulfill all their wishes. In other words, I do not think we can rely on unverified (for safety) agents (even with higher intelligence) to make sure that other agents with higher intelligence are designed to be human-safe. Replacing humanity with something not-human (uploads, augments) and proceeding to ask the replacements the question of how to save humanity is not going to work; at that point, we would already have lost humanity by definition. I am not saying that is not going to happen; it probably will. Most likely, we will see something predicted by Kurzweil (merger of machines and people) (Yampolskiy 2013).

I am also as concerned about digital uploads of human minds as I am about AIs. In the most common case (with an absent body), most typically human feelings (hunger, thirst, tiredness, etc.) will not be preserved, creating a new type of an agent. People are mostly defined by their physiological needs (think of Maslow's pyramid). An entity with no such needs (or with such needs satisfied by virtual/simulated abandoned resources) will not be human and will not want the same things as a human. Someone who is no longer subject to human weaknesses or has relatively limited intelligence may lose all allegiances to humanity because they would no longer be a part of it. So, I guess I define *humanity* as comprising standard/unaltered humans. Anything superior is no longer a human to me, just like we are not first and foremost Neanderthals and only then *Homo sapiens*.

Overall, I do not see a permanent, 100% safe option. We can develop temporary solutions such as confinement or AI safety engineering, but at best this will only delay the full outbreak of problems. We can also get very lucky—maybe constructing AGI turns out to be too difficult or impossible or maybe it is possible but the constructed AI will happen to be human-neutral by chance. Maybe we are less lucky, and an artilect war will take place and prevent development. It is also possible that as more researchers join AI safety research, a realization of danger will result in diminished effort to construct AGI (similar to how perceived dangers of chemical and biological weapons or human cloning have at least temporarily reduced efforts in those fields).

The history of robotics and AI in many ways is also the history of humanity's attempts to control such technologies. From the Golem of Prague to the military robot soldiers of modernity, the debate continues regarding what degree of independence such entities should have and how to make sure that they do not wreak havoc on us, their inventors. Careful analysis of proposals aimed at developing a safe artificially intelligent system leads to a surprising discovery that most such proposals have been analyzed for millennia in the context of theology. God, the original designer of biological robots, faced similar control problems with people, and one can find remarkable parallels between concepts described in religious books and the latest research in AI safety and machine morals. For example, 10 commandments ≈ 3 laws of robotics, Armageddon ≈ singularity, physical world ≈ AI-Box, free will ≈ nondeterministic algorithm, angels ≈ friendly AI, religion ≈ machine ethics, purpose of life ≈ terminal goals, souls ≈ uploads, and so on. However, it is not obvious if god ≈ superintelligence or god ≈ programmer in this metaphor. Depending on how we answer this question, the problem may be even harder compared to what theologians had to deal with for millennia. The problem might be: How do you control God? I am afraid the answer is—we cannot.

REFERENCES

Muehlhauser, Luke and Roman Yampolskiy. July 15, 2013. Roman Yampolskiy on AI Safety Engineering. Machine Intelligence Research Institute. http://intelligence.org/2013/07/15/roman-interview/

Yampolskiy, Roman. September 16, 2013. Welcome to Less Wrong! (5th thread, March 2013). Less Wrong. http://lesswrong. com/lw/h3p/welcome_ to_ less_wrong_5th_thread_march_2013

Index

E

F

Q

R

S

T

U

V

W

Y

Z

Printed in the United States
by Baker & Taylor Publisher Services